The Social Sciences and Democracy

The Social Sciences and Democracy

Edited By

Jeroen Van Bouwel
Ghent University, Belgium

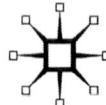

First published 2009 by
PALGRAVE MACMILLAN

Palgrave Macmillan in the UK is an imprint of Macmillan Publishers Limited,
registered in England, company number 785998, of Houndmills, Basingstoke,
Hampshire RG21 6XS.

Palgrave Macmillan in the US is a division of St Martin's Press LLC,
175 Fifth Avenue, New York, NY 10010.

Palgrave Macmillan is the global academic imprint of the above companies
and has companies and representatives throughout the world.

Palgrave® and Macmillan® are registered trademarks in the United States,
the United Kingdom, Europe and other countries

ISBN: 978–0–230–22439–1 hardback

This book is printed on paper suitable for recycling and made from fully
managed and sustained forest sources. Logging, pulping and manufacturing
processes are expected to conform to the environmental regulations of the
country of origin.

A catalogue record for this book is available from the British Library.

A catalog record for this book is available from the Library of Congress.

10 9 8 7 6 5 4 3 2 1
18 17 16 15 14 13 12 11 10 09

Printed and bound in Great Britain by
CPI Antony Rowe, Chippenham and Eastbourne

Contents

Contributors

Patrick Baert is Fellow and Reader in Social Theory at the University of Cambridge. Amongst his publications are *Social Theory in the Twentieth Century* (1998), *Philosophy of the Social Sciences: Towards Pragmatism* (2005) and the edited volume (with B.S. Turner) *Pragmatism and European Social Theory* (2007).

James Bohman holds the Danforth I Chair in the Humanities and is Professor of Philosophy and Professor of International Studies at Saint Louis University. His primary areas of research include political philosophy (deliberative and transnational democracy) and the philosophy of social science (rationality and normativity). He also has done work in German philosophy (Critical Theory and German Idealism). He is director of the Critical Theory Roundtable and co-director of the Philosophy of Social Science Roundtable with Paul Roth and Alison Wylie.

Alban Bouvier teaches the philosophy of social science, the philosophy of social phenomena and theoretical sociology on the one hand, social epistemology and cognitive sociology, on the other hand. He is mainly interested in both the relevance and the limits of methodological individualism and Rational Choice Theory.

Steve Fuller is Professor of Sociology at the University of Warwick. Originally trained in history and philosophy of science, his research programme has been developed in the journal *Social Epistemology* (founded in 1987) and numerous articles and books among which *Social Epistemology* (1988), *Thomas Kuhn: A Philosophical History for Our Times* (2000), *The Governance of Science* (2000), *The New Sociological Imagination* (2006) and *The Knowledge Book: Key Concepts in Philosophy, Science and Culture* (2007).

Helena Mateus Jerónimo is Assistant Professor at Instituto Superior de Economia e Gestão, Technical University of Lisbon, Portugal. Amongst her publications is *Ética e Religião na Sociedade Tecnológica. Os Jesuítas Portugueses e a Revista Brotéria (1985–2000)* (Lisboa: Editorial Notícias, 2002).

Harold Kincaid is Professor and Chair in the Department of Philosophy at the University of Alabama at Birmingham. He is the author of *Philosophical Foundations of the Social Sciences* (1996) and *Individualism and the Unity of Science* (1997), with John Dupre and Alison Wylie, *Value Free Science: Ideals and Illusions* (2007), and with Don Ross, *The Oxford Handbook of the Philosophy of Economics* (2009). He has published numerous articles and book chapters in the philosophy of social science and philosophy of economics.

Philip Mirowski is Carl Koch Professor of Economics and the History and Philosophy of Science, and Fellow of the Reilly Center for Science, Technology and Values, University of Notre Dame. His areas of specialization are in the history and philosophy of economics, with subsidiary areas in evolutionary computational economics, the economics of science, science studies and the history of the natural sciences, and his books include *More Heat than Light* (1989), *Machine Dreams* (2002), *The Effortless Economy of Science?* (2004) and the forthcoming *ScienceMart™*.

Francis Remedios is an independent scholar located in Edmonton (Canada), the author of *Legitimizing Scientific Knowledge: An introduction to Steve Fuller's Social Epistemology* (2003) and many other publications on social epistemology, philosophy of science and epistemology in *History of the Human Sciences, International Studies in the Philosophy of Science, Philosophical Inquiry, Philosophy of the Social Sciences, Ratio* and *Social Epistemology*.

Kristina Rolin is Academy of Finland Research Fellow at Helsinki School of Economics. Her main areas of research are philosophy of science and epistemology, with emphasis on social epistemology and feminist epistemology. She has published articles in *Philosophy of Science, Episteme, Social Epistemology, Perspectives on Science, Science & Education,* and *Hypatia*.

Alan Shipman is Lecturer in Economics at the Open University, England. Amongst his publications are (with M. Shipman) *Knowledge Monopolies: The Academisation of Society* (2006), *The Globalization Myth* (2002) and *Transcending Transaction* (1999).

Stephanie Solomon is a Postdoctoral Fellow in the Research Ethics and Integrity Program at the Michigan Institute for Clinical and Health Research at the University of Michigan. She would like to acknowledge funding from CTSA grant UL1RR024986 and thank Dr. Mark Risjord, for his comments and support.

Stephen P. Turner Is Graduate Research Professor in the Department of Philosophy at the University of South Florida, Tampa. His recent books include *Liberal Democracy 3.0: Civil Society in an Age of Experts* (2003), *The Disobedient Generation: '68ers and the Transformation of Social Theory* (2005, co-edited with Alan Sica), *The Handbook of Philosophy of Anthropology and Sociology* (2007, co-edited with Mark Risjord), and *The SAGE Handbook of Social Science Methodology* (2007, co-edited with William Outhwaite).

Jeroen Van Bouwel is a Postdoctoral Fellow of the Research Foundation (FWO) Belgium, and a member of the *Centre for Logic and Philosophy of Science* at Ghent University. His research interests include social epistemology, the philosophy of the social sciences and general philosophy of science, especially scientific pluralism. His articles have appeared in journals such as *Economics and Philosophy, Philosophy of the Social Sciences, Philosophical*

Explorations, Foundations of Science, History and Theory and *Journal for the Theory of Social Behaviour.*

Erik Weber is Professor in Philosophy of Science at Ghent University (UGent). Most of his research relates to general theories of causation and explanation, or to problems relating to causation and explanation in specific scientific disciplines (psychology, the biomedical sciences and the social sciences). He also works on the metaphysics of causation and on the application of non-classical logics in the philosophy of science.

Where the Epistemic and the Political Meet: An Introduction to the Social Sciences and Democracy

Jeroen Van Bouwel

In his 1937 essay *Traditional and Critical Theory,* Max Horkheimer argued that *traditional theory,* including the heretofore-existing social sciences, had been fixated on the accumulation of facts in specialized, isolated fields of study. Such a fixation had tended to reproduce the existing social order rather than question – let alone challenge – it, Horkheimer contended. In contrast, he proposed a *Critical Theory* that would recognize that the production of knowledge is not to be detached from social power relations and interests, from its embeddedness in society. Like Karl Marx, the younger Horkheimer and his colleagues of the Frankfurt School believed that (critical) theory and knowledge could and should change society by helping those oppressed to identify and emancipate themselves from their oppression. As a theory, it understands the totality of society in its historical specificity, and it is normative, driven by specific social interests, seeking "to liberate human beings from the circumstances that enslave them" (1982, p. 244) and advancing "the abolition of social injustice" (1982, p. 242).[1] This *Critical Theory,* Horkheimer argued, should integrate all the major social sciences, including economics, sociology, history, political science, anthropology and psychology, establishing a unifying new science that elaborates one comprehensive theory, a version of historical materialism.

Overseeing the social sciences at the beginning of the twenty-first century, at least two central aspects of Horkheimer's *Critical Theory* are problematic (see also Bohman 1999). First, the *epistemic* ideal of developing a comprehensive theory unifying the social sciences and its explanatory practice appears difficult in light of the plurality of adequate theories and methodologies developed in the social sciences and the normative endorsement of this plurality by advocates of scientific pluralism. Second, the comprehensive *political* goal of human emancipation corresponding to this single comprehensive theoretical framework does not seem to dovetail with the *smorgasbord* of emancipating, and not seldom conflicting, interests defended by feminists, antiracists, minorities, youth and lesbian, gay, bisexual, and transsexual (LGBT) rights movements, among others. Furthermore, given

1

the lack of epistemic or theoretical unity of the social sciences and its political or practical pendant, the interdependence between the epistemic and the political will have to be reconsidered. An obvious candidate to frame and deal with the plurality of interests encountered in the epistemic and the political as well as their interaction is the idea of *democracy*. It is an idea that plays out on many levels, as we will see in what follows.

The political articulating the epistemic: Improving the social sciences

Many recent contributions to science studies have shown that science is an inherently social process. In philosophy, one can perceive a shift from the traditional 'individual' epistemology to social epistemology; the latter focuses on the social dimensions of knowledge and knowledge production, going beyond individual reasons and causes of belief, evaluating the reliability of social processes in the generation of knowledge and trying to make social requirements for rational scientific inquiry explicit (which does not imply that science is merely a social construction, just that considering the social dimensions of knowledge and knowledge production is necessary). The attention for the social aspects of scientific inquiry helps us to depart from an image of the ahistorical, acontextual, autonomous, uniform and interchangeable knower and to take into account the differences qua personal history, experience, expertise, concerns, interests, values, context and so on among knowers; in short, there is more to scientific inquiry than The Scientist following The Scientific Method.

The focus on the social aspects and the diversity of knowers and their perspectives, interests and values, also brings the political to the fore. Is science in the common interest of all, a common good (whatever that means), or are some perspectives and interests better served than others by science? These questions lead to more questions: How are different epistemic and nonepistemic interests addressed by the social scientific disciplines (or how could they be addressed)? Is the existing plurality of theories, methodologies and forms of explanation in social science due to the different interests addressed, or due to other sources of plurality, such as the complexity of the world? Discussing the existing plurality, we could as well ask whether it is desirable and should be endorsed – as done by *scientific pluralism* in contemporary philosophy of science – or whether the plurality is merely temporary and monism should be our goal.

These questions concerning the theoretical and methodological plurality in the social sciences, as well as the plurality of interests and values, provide a clear opening for the political, and in particular the idea of democracy, on at least two levels: First, could we not deal with the plurality of theories and perspectives in social science by framing this plurality within a democratic framework? Would that not help us to clarify and manage the relations among different perspectives, for instance, the coexistence of orthodox and heterodox theories? Can we draw fruitful parallels between *models of science* – dealing with a plurality of epistemic interests – and *models of democracy* – dealing with a plurality of political, social, economic and moral interests?

Second, does the plurality of different interests and values involved undermine the impartiality and political neutrality of social science? Or can, on the contrary, a democratic inclusion of these interests and values improve the epistemic, the social sciences? The impartiality and political neutrality are presumed by the image of the social scientist as the *technocrat*. The technocrat is a social scientist – modelled on the engineer – that provides technical insight and optimal problem-solving strategies to the public and society and is impartial vis-à-vis the ultimate goals the public and society should pursue. These goals should be decided upon by the public in the electoral process and by elected politicians.

Besides the technocrat, one can distinguish two more types of social scientists in their relation with the public and society; let us label the first type as the *epistocrat* or *expertocrat* and the second as the *democrat*, a discussant or participant in a dialogue. The view of the social scientist as an epistocrat is, for instance, advanced by Auguste Comte. Taking sociology serious as a science, Comte considered it imperative to bring political life in line with scientific sociological truth, instead of basing it on decisions made by the public; public discussion could be abolished in favour of expert rule. Thus, the epistocrat not only provides problem-solving strategies like the technocrat does, but also knows the goals that society should pursue.

Both the technocratic and the epistocratic view on social science might find fewer advocates nowadays, not only because theoretical and methodological pluralism result in contradictory or conflicting prescriptions and advice to policy makers, which seems hard to square with the image of social scientist as the optimal problem-solver, but also because the role of values and the partiality of 'depoliticization' are more and more acknowledged. Critical historians of the social sciences have shown the existence of imperialist, racist and sexist tendencies in social scientific research, for instance, in the anthropological interpretations of non-Western people, the exclusion of women's voices in the economic conceptions of labour and household and via all kinds of naturalisations of social inequality.

These observations led many to rethink the epistemic relation of social scientists with the public and introduce the type of the social scientist known as *democrat*. According to this view, the impartiality and universality claimed by certain social scientists (usually characterised as Western, white, middle-aged and male) to the exclusion of (and irrelevance for) other perspectives should be replaced by an inclusionary and democratic approach – which might involve nonscientific stakeholders or lay scientists providing insider, local and/or lay expertise – in order to obtain better social science. (One can, for instance, think of communities drawing on centuries of tested local knowledge, beliefs and practices found over time to be critical to community survival, reproduction and protection of resources.)

But, what should such an inclusionary and democratic approach look like? It raises questions like: How and to what extent should the social scientists engage in a dialogue with 'outsiders'? Can they play a role in decisions concerning (a) the research agenda and topics to be funded, (b) the use of the research results, and/or (c) the actual research process and the epistemic justification of scientific knowledge? Traditionally, (a) and (b) have been less controversial than (c); but, to what extent can 'outsiders' appropriately challenge scientists' assumptions and participate in scientific debates? A central challenge in democratizing social science seems to be to find the right balance and division of labour between what should be delegated to scientific experts (and considered in an epistocratic way) and what should be kept within the sphere of public discussion (and considered democratically).

This democratisation of social science and the inclusion it involves can either be motivated by *epistemic* reasons, that is, democratisation understood as epistemically necessary to obtain an improved, more adequate social science, or, be motivated by *political* reasons, that is, understood as politically more just and corresponding a democratic society. Such a society cannot support an imperialist, racist, sexist, or another prejudicial social science that risks to consolidate oppression and inequality.

The epistemic articulating the political: Improving democratic theory and practice

In the previous section, we drew on the political structure of democracy to clarify and improve science and scientific expertise. Equally interesting is exploring the other direction, for instance, analysing whether the social interaction among scientists (described in social epistemology) might be exemplary for the functioning of a democratic society. The variety of social epistemologies (e.g., consequentialist and procedural variants) can, furthermore, clarify the different articulations of democratic decision-making (e.g., rational and pure proceduralism, cf. Peter 2007) – which brings us to democratic theory.

Democracy as a practice needs ongoing improvement and adaptation to new developments in order to achieve its aims, such as, *inter alia*, avoiding the great harms of political power, war and famine. The reflection on democratic practice and the elaboration of democratic theories is traditionally part of social science. It has generated a plurality of democratic theories, sometimes classified as procedural, constitutional or deliberative; other times as representative, direct, participative and deliberative democracy, radical pluralism, democratic pragmatism, and so on (see also Cunningham 2002). We will not go into the minutiae of the debates here; it suffices to say that democratic practice might benefit from these debates as well as from the variety of democratic theories. That said, democratic theory can still be improved – as will be argued later on in the book – for instance, by scrutinizing the social assumptions of democratic normative theory and the value assumptions of social scientific studies of democracy.

Furthermore, the social sciences also help us to tackle the obstacles that democracy as a practice must overcome. Identifying these problems (e.g., globalisation) incites practical reinterpretations of democracy that address such problems or guides us to improve democracy by tackling tendencies that might threaten democracy (like failures of rationality in decision-making) by modifying the democratic environment so that such tendencies cannot persist unreflectively.

Finally, the social sciences are often regarded as crucial to the functioning of democratic societies, as being part of the democratic process, providing adequate knowledge and informing the public in the democratic society. This brings us back to the distinction introduced above between the social scientist as technocrat, epistocrat or democrat. The question here is not so much to what extent the social scientist should take into account or consult the public and its knowledge, but rather what role the social scientist should play in a democracy – as a technocrat, epistocrat or democrat – and what status should be ascribed to social scientific knowledge (in comparison or competition with the public discussion) in the democratic process. The social sciences might obstruct or facilitate a democratic society, just as a democratic society can be an obstacle or a facilitator for the social sciences. This ambivalence waits to be tackled.

Materializing the continuing dialogue between the epistemic and the political

As put forward in the previous sections, the epistemic practice of the social sciences can benefit from an analysis in political terms, in particular using the concept of democracy, just as well as political practice can be improved by an input from the social sciences. In order to endorse this interaction between the epistemic and the political, and in order to implement and develop democracy on the different topics identified above, an appropriate *governance of social science* is vital. This should facilitate and lubricate the relation between the social sciences and democracy besides handling possible obstacles.

Having considered the theoretical and methodological plurality in the social sciences as well as the plurality of interests and values in the first section, one perceives immediately some problems that a democratic science policy would have to deal with. If different interests are being addressed by different scientific perspectives and theories, the question arises whether the current science policies are adequate for democracies; whether all interests present in society are being served. Do current policies presuppose social science to be in the interest of all? Should governments continue to support the social sciences if they are partial – making 'political' pronouncements – with the risk of engendering a state-sponsored source of opinion within public discussion? Would that not violate the political neutrality of the state? Can we develop an appropriately democratic science policy for the social sciences?

Scrutinizing the possibility of a *democratic* science policy, at the outset one should discuss what kind of democracy should be preferred (direct, representational or any of the other varieties of democracy mentioned in the previous section). Furthermore, it has to be decided which issues should be subjected to democratic decision-making: the research agenda setting; the applications of the research output, that is, scientific knowledge; and/or the actual scientific research process – regulating science and scientists. The latter would require the development of democratic governance-of-science legislation – some form of social contract – to govern the community of scientists, with science as a social institution accountable to the democratic state. What could such a social contract look like? Would it necessarily facilitate the relation between the social sciences and democracy and advance the democratization of social science?

The above contractarian approach is opposed to the invisible hand account of science, considering science as the marketplace of ideas. The idea of the invisible hand has been used by several contemporary philosophers of science (e.g. Philip Kitcher and Alvin Goldman), and it is a good example of using an *economic* model to comprehend scientific activity – aspiring to unveil the logic of science. The economics of science and knowledge (a booming industry in times of knowledge economies and information societies) is but one way of studying the process of science. Given that scientific practice can be considered as a social process, all of the *social sciences* can in principle provide us with conceptual tools to analyse science (another being the use of *democratic* models elaborated within *political science*, cf. supra). Notwithstanding the often-contradictory results of these social studies of science, understanding the dynamics of science – thus discussing the social science of social science – seems indispensable in order to develop an adequate science policy and efficaciously govern science.

Considering the institutional and economic context of science, many scholars have been pointing at the growing commercialisation of science. Since 1980, a broad array of innovations has caused a profound reorganisation of the university and of the structure of science. A first question one might raise is to what extent this changing institutional context of science disproportionally serves the interests of some groups. For instance, in pharmaceutical research, burying certain negative results that risk harming commercial interests is not unusual. Moreover, many of the commercially funded researchers work with contracts which give the funder say over what is published. Second, one might wonder how commercialisation affects the research agenda – skewing the agenda towards patentable research, industrial applications and other knowledge production with a high economic value. The epistemic impact of this transformation of scientific knowledge from a public good to a positional good might be enormous, therefore, it is imperative to understand how science is funded and organized, for this affects which science is produced. How the commercialisation acts on the social sciences, and their opportunities to democratize, will also have to be taken into account.

The commercialisation might as well drastically change the power balance between scientific disciplines, not only within the social sciences, but as well between the social sciences and the natural sciences (biology, for instance) – with the latter colonizing the social sciences. Is it desirable that the social sciences keep their sense of autonomy as a body of knowledge distinguishable from the natural sciences and the humanities? On what would their distinctness be based and how could it be assured? Would they have any special relationship with specific groups in society? Could they be democratic *par excellence*?

This book: A tour d'horizon

The relation between the social sciences and democracy has many facets – glimpses of which were caught in what preceded. The contributions to this book will elaborate on these different facets by presenting concrete cases, clarifying terminology, adding complexity and hopefully being thought provoking. We will now take the reader on a *tour d'horizon* of the book, introducing the different parts and chapters (inevitably falling short of capturing the richness of these different contributions).

Part I of the book deals with the relation between social scientific experts and the public in a democratic society. What role should the social scientist assume: the one of the technocrat, the epistocrat or the democrat (with the latter being subject to many different interpretations)? In Chapter 1, Patrick Baert, Helena Mateus Jerónimo and Alan Shipman tell the story of the technocratic model – and its link to the social sciences' struggle for identity – to find it foundered in our time. Considering the social scientist's possible contributions to public participation and debates concerning technological and scientific management and decision-making, the authors explore the ways of engaging technocracy in dialogue, emphasising the potential of dialogical social science and its capacity to broaden and assist democratic practice.

The tension between expertise (as provided by the technocrat or the epistocrat) and a democratic dialogue plays a central role in Chapter 2 as well. Stephanie Solomon scrutinizes the call for democratizing science, a call motivated by the history of the social sciences propagating unjust politics and espousing biased (Western) knowledge as universal. Discussing proposals to democratize science from feminist theory (Lynn Hankinson Nelson), sociology of science (Brian Wynne) and the practical social science approach of community-based research, Solomon analyzes whether these attempts succeed in maintaining an epistemically coherent notion of expertise in science. Subsequently she examines how to combine the ideals of expertise and the ideals of democracy in a single idea of democratizing science – incorporating nonscientists in the social scientific discussion.

Whether, and how, to include nonscientists or outsiders in particular communities preoccupies Kristina Rolin too. In Chapter 3, she develops a stakeholder theory of scientific knowledge starting from the question whether stake-holding outsiders have a role to play in epistemic justification and – if they have – what this role might be. Philosophers of science have traditionally acknowledged that those outsiders can have a say in decisions about the research agenda or the end use of scientific knowledge, but they have been assumed to lack authority in issues of epistemic justification, a position Rolin argues against. She not only clarifies the role stakeholders can play in scientific debates, but also the epistemic responsibilities of scientists vis-à-vis stakeholders, that is, their duty to engage in scientific debates with stakeholders under certain conditions. Rolin's, Solomon's and the first chapter propose different balancing acts concerning the exact input of the social scientist and her relation with the public, that is, different doses qua technocracy, epistocracy and democracy, a boon to the discussion.

Part II discusses the ways in which the social sciences can help to improve democracy, both in theory and in practice. In Chapter 4, James Bohman analyses how the social scientific study of democracy can become one aspect of a practical theory or praxeology directed to improving democratic practice. Improvements can be suggested by understanding and explaining how democratic institutions promote preferred outcomes, like the avoidance of famine (cf. Sen's hypothesis that there has never been a famine in a democracy) and war (cf. the democratic peace hypothesis that democracies do not go to war with other democracies). Another way in which the social sciences can contribute is by clearly identifying the obstacles to democracy. One example could be globalisation – we should know what globalisation is exactly in order to be able to discuss an adequate democratic reform. Another one deals with human reasoning: is there an inherent tendency of human reasoning to be systematically mistaken that undermines democratic deliberation? If so, how can this be remedied? Discussing these examples, Bohman articulates what form of social science and democracy is required for the realization of this praxeology oriented to improving democratic practice.

In Chapter 5, Harold Kincaid explores the interactions between normative democratic theory, the social sciences, and the philosophy of science. He scrutinizes whether they do not share common assumptions which are mistaken, namely, that practices can be explained and evaluated by identifying the formal procedures that are followed, that these procedures can be identified and understood independently of their social embodiment and the sociological processes at play and that these procedures can be so understood in a value-neutral way. Kincaid discusses the social assumptions of democratic normative theory, the value assumptions of social scientific studies of democracy and social assumptions and democratic norms for science, respectively. One could question, for instance, whether many of the assumptions of normative democratic theory are actually consistent with solid findings in the social sciences. Kincaid argues that a thicker notion of the social will contribute to improvements in all three fields, and eventually in democratic theory and practice.

Where Parts I and II focus on the interaction between social science and society, the contributions to Part III concentrate on science itself and how to understand its dynamics. Models of democracy can be a fruitful source for modelling scientific practice, or so it will be argued. Relying on the parallels between models of democracy and models of science, Chapter 6 (mine) questions the ideal of the scientific consensus. Science and its consensus ideal are often understood as exemplary for deliberative democracy. Starting from Chantal Mouffe's critique on the consensus ideal of deliberative theories in democratic theory, the consensus ideal in science is questioned and the value of dissent articulated. Mouffe's model of democracy, labelled agonistic pluralism, is advanced as a model of social science, endorsing a plurality of theories and perspectives in social science as well as clarifying a framework for understanding and managing the relations among different perspectives, for instance, the coexistence of orthodox and heterodox theories in economics.

Where Chapter 6 focuses mainly on the interaction between different perspectives and approaches, Chapter 7 concentrates on the ethical and political aspects of the dynamics within one approach (i.e., within a research programme, paradigm or school). Alban Bouvier explores to what extent contractualist models of groups like Margaret Gilbert's or Philip Pettit's can help us to understand the dynamics of scientific groups, bolstering his conceptual analysis with case studies from sociology (the French School), economics (the Austrian School) and the early history of quantum mechanics. His analysis offers political philosophical concepts to catch the degree of liberty and democracy within these groups in relation to collaboration and collective deliberations leading to the modification of certain principles of a school or research programme. Furthermore, applying the distinction between positive and negative liberty as well as Pettit's idea of liberty as absence of domination in relation to these collaborations, Bouvier suggests a normative guideline in the conduct of science. Here as well, it is shown how political philosophical concepts can help us to model science.

These two chapters contribute to a thorough understanding of the dynamics of science, an important aspect to take into consideration in designing an efficacious science policy and a democratic governance of science, the theme of Part IV. In Chapter 8, Stephen Turner questions whether the liberal idea of political neutrality, namely that democracy requires a neutral state that does not take sides in public discussion, poses a problem for sociology. If one understands sociology to be a contributor to political discussion, then its funding by the state seems unallowable. Turner discusses in particular public sociology which is intended to have a political impact, to give voice and support to particular movements and groups, and which seems to violate the dictum of political neutrality. Turner analyses Michael Burawoy's justification of public sociology looking for a way to overcome this difficulty, whether social sciences might be partial and neutral at the same time. Moreover, this chapter offers us an insight into the possibilities of a dialogical rather than an expert relation between social science and society, as intended by public sociology.

The state subsidization of social science does not only have to deal with questions of political neutrality, but also with questions of priorities and ways of – democratically or not – distributing the limited resources. In Chapter 9, Erik Weber scrutinizes the proposals made by Philip Kitcher concerning a democratic science policy. Kitcher distinguishes internal elitism, external elitism, vulgar democracy and enlightened democracy as possible forms of science policy, and he advocates enlightened democracy. Weber wonders whether Kitcher's arguments in favour of enlightened democracy and against the other forms of science policy are sound. Should he not make a distinction between direct and representative democracy in rejecting vulgar democracy? Do Kitcher's arguments against external and internal elitism eliminate the option of elitism completely? What conclusions for a democratic science policy can we draw after having revisited Kitcher's proposals? Can scientists themselves decide where the research money granted by the state will be spent?

Discussing the distribution of research money brings us back to the economics of science, in which science is usually analysed as a marketplace of ideas run by the invisible hand. However, some economists oppose this orthodoxy, as Philip Mirowski elaborates in Chapter 10, discussing the work of Richard Nelson, Sidney Winter, Paul David, Giovanni Dosi, Benjamin Coriat, Paul Nightingale and others – dubbing them the *3E* school (*Evolutionary Economic Epistemologists*). They are appalled by the neoliberal turn taken by the neoclassical orthodoxy since roughly 1980 and consider their "new economics of science" to be a defender of the virtue of science against neoliberal and other modern privateers of science and knowledge. Notwithstanding their intentions, Mirowski notices that many 3E figures straddle a sequence of intolerable contradictions and appear to backslide into dependence upon the marketplace of ideas when studying science. At the end of the day, 3E does not seem to be so different from earlier conventional doctrines of the neoclassical economics of science (e.g., exiling power to an unexplained residual, the complete commodification of knowledge, failing to really make use of history in discussing technological change and the operation of science). Worse, some 3E protagonists, while pleading the opposite, succeed in granting legitimacy to the neoliberal approach to the marketplace of ideas well beyond the circle of original neoliberal econo-mists. More positively, Mirowski does see other 3E figures, especially the European wing, doing better.

Mirowski's analysis is not only important in relation to the understand-ing of the dynamics of science and the design of an adequate democratic science policy, it also lays bare a worrying evolution, namely, that the study of science itself – at least, the economics of science – might become co-opted to the modern neoliberal regime of globalized privatization of science given that the neoliberal ideas are legitimized by and ingrained in the writings of many 3E scholars, self-declared opponents of privatization. This would turn the social studies of science themselves into obstacles to the democratiza-tion of social science.

In Part V of this book, two more obstacles to the social sciences and dem-ocracy will be scrutinized. In Chapter 11, Francis Remedios discusses what is at stake in the commercialisation of scientific knowledge. As argued above, commercialisation might have an enormous epistemic impact and seems hard to square with the democratisation of science – the latter requiring more than commercial interests to be taken into account. Remedios ana-lyses how commercialisation is interpreted and evaluated in the work of Philip Mirowski (focusing on the rise of neoliberal doctrines in the postwar era) and Steve Fuller (advocating a republican approach to the governance of science), and he compares their approaches, identifying convergences and divergences between both and considering to what extent they suggest a better regime for science.

In the last chapter of this book, Chapter 12, Steve Fuller discusses the autonomy of the social sciences as a body of knowledge distinguishable from, on the one hand, the humanities and, on the other, the natural sci-ences. He explores the theological vestiges of social science, that is, the spe-cial treatment given to humans vis-à-vis all other creatures both in terms of the values ascribed to human things and their modes of study (contra nat-ural science), as well as the equal eligibility of all, not simply elite, humans to such treatment (contra the classical humanities), granting a central place to John Duns Scotus. Fuller argues that Duns Scotus put the metaphysical framework in place to engage *Homo sapiens* in *humanity* as a collective project of self-transformation, stressing our world-making capacities and achievable perfection – the source of modern notions of progress. The unique equality that humans enjoy as having been created *in imago dei* provided the historic ontological and epistemological underpinning for democratic politics and the democratisation of social life more generally.

Fuller sees the distinctness of the social sciences, which seemed so sali-ent over the last three centuries, disappearing today. And with the loss of the social sciences' distinctness, the terms of democracy are equally up for renegotiation. On the one hand, normative categories traditionally confined to humans, especially legal ones pertaining to rights, are being extended to animals and even machines. On the other hand, there are increasing attempts to withhold or attenuate the application of such normative categories to, say, the disabled, simply the unwanted or unproductive humans. The scientific and political question should then be what is worth continuing to defend as distinctly *human*, according to Fuller. The idea of *humanity*, and its future, augurs the fate of the social sciences and democracy.

Meeting the epistemic and the political in our time

When Max Horkheimer wrote his *Traditional and Critical Theory* in 1937, he deplored the lack of normative guidelines coming from the social sciences to change the social order as well as the presumption that the production of knowledge would be detached from social power relations and interests. Looking at the broad field of science studies nowadays, two big camps can be distinguished – one can be labelled *social studies of science* and the other *philosophy of science* and/or *analytical social epistemology*. Surveying the camps, we can see that Horkheimer's analysis still holds: on the one hand, most of the social studies of science, while articulating the power relations and social interests at play, keep a distance from formulating normative guidelines or theories to change science and seem to prefer celebrating contingency. On the other hand, most philosophy of science contributions, while not recoiling from formulating normative guidelines, neglect power relations and social interests.

Through bringing together scholars from both camps analysing the interplay between the epistemic and the political, this book is an attempt to overcome the division in camps – hoping that the normative approach of the one can be combined with the attention for power relations and social interests of the other. In times of systemic crisis (fortunately not of a magnitude of 1937, yet), it is recommended to revisit Horkheimer – be it in a more democratic outfit appropriate to our more democratic societies – and take his critique of the social sciences seriously: trying to elaborate a more normative stance towards the direction the social sciences should take (more so than most social studies of science) and taking the embeddedness of social science in society into consideration (more so than most philosophers of science), so that we can scientifically question and democratically change the direction our society is heading.[2]

Notes

1. The German version of the latter is actually more telling: 'das mit ihr selbst verknüpfte Interesse an der Aufhebung des gesellschaftlichen Unrechts.' (Horkheimer 1970, p. 56).
2. I would like to thank all contributors for their meticulous preparation of manuscripts, their patience and their cooperation, as well as Jan De Winter,

Petri Ylikoski, Rogier De Langhe and Linnéa Arvidsson for having read parts of earlier versions of this book and provided valuable feedback on the content and on the presentation. Finally, as always, many thanks to my colleagues at the Centre for Logic and Philosophy of Science (Ghent University) and the Research Foundation (FWO) – Flanders for making all of this possible.

References

Bohman, J. (1999) 'Theories, Practices, and Pluralism. A Pragmatic Interpretation of Critical Social Science.' *Philosophy of the Social Sciences* 29(4), 459–80.

Cunningham, F. (2002) *Theories of Democracy: A Critical Introduction*. London: Routledge.

Horkheimer, M. (1970) *Traditionelle und Kritische Theorie* (Orig. 1937). Frankfurt am Main: Fischer Bücherei.

Horkheimer, M. (1982) *Critical Theory: Selected Essays*. New York: Continuum.

Peter, F. (2007) 'Democratic Legitimacy and Proceduralist Social Epistemology.' *Politics, Philosophy, and Economics* 6(3), 329–53.

Part I

Democratizing the Social Sciences: Balancing Expertise and Dialogue

1

Social Sciences and the Democratic Ideal: From Technocracy to Dialogue

Patrick Baert, Helena Mateus Jerónimo and Alan Shipman

Social sciences were launched on a wave of expectation that scientific study of society would yield practical benefits, through improved understanding, policy and organisation. This expectation was strengthened through the twentieth century: social sciences staked out new disciplinary areas and developed distinctive methodologies (such as game theory) alongside those imported from the natural sciences. Promising transferable skills for students as well as technical insights for policy makers, they were major beneficiaries of universities' postwar expansion. But the social sciences now have a harder time selling themselves, partly because they lack any clear consensus as to the values they serve and the direction they should take. Social scientists are a dispersed group with clearly different views about what they stand for, why we should read them, and why universities and governments should continue supporting them.

Despite their differences, most social scientists agree on a fundamental principle underlying social research. It is the idea that the social sciences are somehow crucial to the workings of democracy – that they are part and parcel of the democratic process. Natural sciences, notably physics, were widely credited with rescuing democracy from mid-century threats, through superior technology and associated project management. Social sciences made an equally powerful claim to be 'winning the peace', by showing how long-running problems of economic instability, poverty, inequality and injustice could be solved by efficient policy action without eroding individual and commercial freedom.

This form of justification for social research was advanced by social scientists – not social theorists. Social theorists also see themselves as inextricably linked to the democratic tradition, while citing a very different connection. Whether through a theory of communicative action or a renewed pragmatist tradition, they profess to provide the philosophical building blocks for the creation or recovery of a truly democratic society (e.g., Habermas 1981a, 1981b, Bernstein 1992). Theory is regarded as essential to the making of democracy, and in the process, the very act of theorising is justified. It is no longer an idle, purely academic endeavour; it is directed towards making this world a better place.

Social scientists see themselves equally connected to the democratic tradition, but in a different, and distinctly more modest, fashion. Their professional self-image is basically as accomplices to democratic policy implementation. Through the electoral process, citizens decide on the ultimate goals that will be pursued, with politicians embodying or personalising those goals. The social sciences provide technical insight into which means might be most effective in pursuing the goals that are given. It is essentially a technocratic view of the social sciences in that they are neutral vis-à-vis the ultimate goals that are being pursued.

In their writings, Max Weber and Karl Popper have articulated and defended this technocratic view. Weber (1949, pp. 1–47) was anxious to distance himself from the Marxist tendency to blur what is and what ought to be. He insisted that we cannot infer ultimate goals from our social research, just as our values should not interfere with our research. In a similar fashion, Popper (1971a, 1971b, 1991) was particularly keen to distance himself from grand-scale social changes and to promote piecemeal engineering in which the social sciences would occupy a crucial role. The social sciences would enable us to act with caution and to change things incrementally and responsibly. Popper's opposition to the diachronic 'big picture' of historicist grand theorists (like Marx) was complemented by Friedrich Hayek's opposition to the synchronic 'big picture' of political central planners. The dispersion of knowledge among individuals meant, in the 'Austrian' view, that experts could never build up a comprehensive picture capable of centrally coordinating society (Hayek 1945); and that social scientists could never achieve the separation between objective fact and subjective value that natural scientists had credibly claimed. "The reason for this is that the object or the 'facts' of the social sciences are also opinions – not opinions of the student of the social phenomena, of course, but opinions of those whose actions produce the object of the social scientist" (Hayek 1979 [1952], p. 47).

The technocratic view implies a homology between the workings of social science and the processes of democracy. In their ideal-typical forms, both operate according to procedures of transparency and openness. The works of social scientists, like other scientists, are open to empirical scrutiny, and hypotheses will be adjusted if necessary. Likewise, transparency and procedures of unrestrained critique and defence are supposedly endemic to the principle of liberal democracy, even if reality does not quite match those lofty principles (Elster 1998). It is as if there is an elective affinity between social research and democracy, both operating according to similar principles.

But the technocratic view also allocates a distinctly secondary role to the social sciences in two ways. First, no longer stipulating which ultimate values ought to be pursued, social research has become a handmaiden to the political process. In contrast with earlier holistic approaches aimed at macroscopic changes, the technocratic model no longer implies a heroic picture of social research, but it allows us to make informed policy decisions and to avoid catastrophic mistakes. Social research sets out how to get most effectively where we want to go and points out the possible negative, unintended consequences of some of the possible routes we might wish to take. Second, if natural sciences prove more appropriate than social sciences for showing democratic governments how to achieve their goals, then social sciences will be pushed to the sidelines. This may happen if, for example, medical researchers and epidemiologists assert themselves as the best judges of where health expenditure should be allocated, and atmospheric chemists or meteorologists become viewed as the best source of solutions to climate change.

Social sciences' struggle for professional identity

Professions often develop a clearly articulated self-image. If they do, they invariably present themselves to themselves and to others as providing a service to the larger community in which they find themselves. This self-image suggests that the members of the profession are involved in activities which ultimately serve central values of society. They are doing something useful (MacDonald 1995, pp. 187–209), although their usefulness is measured more widely than the productivity or value-added measures applied to nonprofessional workers.

Entry qualifications for professions and rules of professional conduct have traditionally allowed professionals to define their status according to social values rather than economic value. As classically defined (e.g., Perkin 1989, chapter 1) professionals must undergo years of formal training, receive appropriate qualifications and thereby gain membership of an institution that certifies their right to practice. They are bound by rules set by their professional institution which override those of the private company or public agency they work for. The qualification process limits professional numbers and enables the institution to set comfortably high fees, avoiding the competition that drives down pay in occupations where entry is not restricted. Professional rewards are also boosted by the belief of their clients that higher payment gives incentives for better performance and by clients' inability to gauge a professional's worth *ex ante*, which means that quality of service is inferred from the fee charged (Akerlof and Yellen 1986). With these financial assurances, 'classic' professions can deny any concerns about money and focus on the social and public benefits of their work.

Some professions spend less time than others developing or defending their self-image. A coherent and persuasive self-image is less important for those professions for which there is already a broad consensus as to their viability and effectiveness. The medical profession, for one, does not need to convince us about its value. We generally accept that they are doing something important and useful, even if we have qualms about a specific doctor or treatment or about the general paradigm within which the current medical establishment operates. We accept that they are working for the best interests of society. Generally speaking, they get things right; and although mistakes occur, this can best be determined by other experts within the profession. So, older professions are trusted to be self-regulating, setting and enforcing their own standards of valid knowledge and good conduct. Engineers, lawyers and actuaries are among the professions that have generally achieved and maintained this status, based on public acceptance of their demonstrated value.

But for some professions, there is not such a broad recognition that they are doing something valuable. Academics spring to mind, and amongst academics, social scientists are very much a case in point (Flyvbjerg 2001, pp. 1–9). Social scientists are regularly attacked or derided for their perceived lack of societal value. They are frequently mocked for articulating what we already know (and, of course, expressing it in an opaque language) or for studying trivialities. Alternatively, they are criticised for dressing up political slogans as academic truths. To the extent that there has been a questioning of trust in professions in general (O'Neill 2002), social scientists have been accused of promoting an assault on the legitimacy of other professions, something which does little to enhance their own (Sokal 1996a, 1996b; Gross and Levitt 1997; Ashman and Baringer 2001). Finally, social scientists are sometimes portrayed as exercising excessive power, thereby harming society because they lack the outcome or procedural effectiveness of older professions. So social scientists' contribution to inflationary bubbles, town-planning disasters and sexist classroom practices (Shipman 1997) are recalled far more easily than any good they may have done.

It is therefore not surprising that the social scientists spend a lot of energy formulating and fine-tuning a positive self-image. We see them often defending what they do and why they do it, explaining how their activities are ultimately for the greater good. Obviously, not all social scientists are involved in justifying their activities to such an extent, and some disciplines do so more than others (contemporary sociology does it more than economics, for instance). But social scientists certainly justify their existence more than most other professions and, crucially, more than their colleagues in the natural sciences. This form of meta-theorising – legitimising their own activities – is ubiquitous in the social sciences.

Technocratic legitimation and limitation

Under pressure to justify their existence (and employment often at public expense), professions do not just explain why their activities are worthwhile;

they also set clear boundaries and outline why others cannot achieve what they manage to do. Once social scientists are seen as providing the intellectual input for social engineering, the question arises what sets them apart from other social commentators. According to the technocratic view, it is both the notion of expertise and a distinct methodology that makes the social sciences so well placed to assist the democratic process. Indeed, the institutionalisation of the social sciences was, in most countries, accompanied by a clear focus on methodological issues that would distinguish scientific accounts on the one hand and speculative or philosophical attempts on the other. In the course of the twentieth century, methodology became a central component of the professional training process.

Awareness of the limitations of this technocratic view has grown in two ways. First, we now realise that its implied fact-value distinction is not as clear-cut as its advocates first claimed. We have become accustomed to the notion of methodological pluralism in the social sciences, so we have become aware of the extent to which even the most rigorous empirical research can be contested on methodological grounds. There are generally competing theories and contrasting prescriptions, even after increasingly comprehensive data sets and complex statistical techniques have been used to try to identify and calibrate a unique explanatory model. Social sciences have experienced great difficulty in generating clear advice for policy makers that is both supported by evidence and commands a consensus across the profession. If social sciences' contribution to democracy consists solely in its ability to furnish policy makers with workable solutions for specific problems and measurements of policy effectiveness, then expectations of their contribution have been disappointed.

Second, we have become more sensitive to the way in which politicians and civil servants can use and misuse social research information to defend their actions, in some cases leading to the use of 'counter-findings' by other political groups (see also Lash, Szerszynski and Wynne 1992). We have become warier of social research being used as a tool to justify actions rather than a means for achieving democratically prescribed values. Natural scientists were forced to become more involved in shaping the use of their findings as the destructive potential of discoveries in physics, chemistry and biology increased. As the social sciences have tended to seek explicitly a political application of their work, they have been equally compelled to consider the possible misrepresentation and misapplication of their conclusions.

Greater scepticism towards the credibility of technocratic advice has been encouraged by internal as well as external critiques of the natural sciences. If the validity of the natural sciences is in question, then *a fortiori* the same applies to the social sciences. Natural scientists have recognised that even their best-supported explanations are only provisional, and the latest in a long line of models that once appeared authoritative but were subsequently rejected as falsified (Stanford 2006, pp. 6–9). In line with insights by philosophers like Quine (1975) and Kukla (1996), they have conceded that two or more competing explanations might fit the evidence equally well, and that these 'empirically equivalent' theories might lead to different expectations and prescriptions. Natural scientists have been shown, in historical and sociological studies, to arrive at their concepts of truth and appropriate policy by routes very different from those of abstractly specified 'scientific method' (Kuhn 1970, Latour and Woolgar 1986). Likewise, the extent to which professional practice is based on 'objective' knowledge has been challenged historically and theoretically. For example, medicine is alleged to have biased its beliefs towards the effectiveness of treatments by favouring positive over negative findings for publication (Turner et al. 2008), and to have relied far more on 'placebo' effects than its practitioners ever acknowledged (Porter 1997). It has also been argued that the superior performance that some managers ascribed to application of engineering and mathematical techniques may actually have been due to random fluctuation (Malkiel 2007; Taleb 2001).

Even when social scientists could converge on an agreed explanation for phenomena with associated policy prescriptions, this raises alarm among many advocates of democracy and pluralism. Their fear was that a successful partnership between social sciences and technocrats would 'depoliticise' large areas of social organisation, ruling them out of public debate and imposing one-best-way regardless of democratic preference. Before the global financial-market breakdown of 2007–9, 'free-market' economic policy prescriptions were observed to have been taken up by governments of all political stripes, even those that had opposed such economic liberalism. Economists had achieved widespread acceptance that full employment, innovation and sustained growth required balanced budgets, private ownership, deregulated financial markets, 'flexible' labour markets and removal of trade and capital barriers. So whichever party took power, promising greater prosperity, these were the expert solutions they were urged to adopt. This meant that, for a long time, political debate was minimised with regard to one of the central areas of public policy.

Similarly, organisation studies had delivered the harsh 'scientific' truth that efficiency required people to work in large corporate or public-service bureaucracies, shop in supermarkets and live in high-rise apartments, regardless of how strongly they would have preferred a small-company, suburban corner-store life. And 'evidence-based medicine' had persuaded health services to abandon folk remedies and 'holistic' treatments to which people showed great attachment, but science could ascribe no measurable effect. From the early twentieth century onwards, fears that technocracy would stifle political debate were expressed both on the political left (e.g., Galbraith 1967) and the right (e.g., Dahl 1998). Weber's (1968) fears that bureaucrats' expert neutrality could give them excessive power that could be amorally directed, insensitively to intended missions have been echoed by later historians and sociologists (Mommsen 1974; Bauman 1991) and by economists concerned that they will stifle entrepreneurial as well as social freedom and creativity (Schumpeter 1942). From the 1980s, the political left bemoaned the erosion of its social policy hopes by market-driven economic policy constraints; but 50 years earlier, the political right had protested against similar erosion as the logic of planning, by large corporations and government, threatened to override entrepreneurial freedom. In short, whilst the social sciences have often seen themselves in a technocratic light, there has also been a growing uneasiness within society – and indeed within the community of social scientists itself – about the uncontested acceptance of technocratic solutions.

Social sciences and policy: Promoting audience participation

While the social sciences fluctuate uneasily between working to contribute to technocracy and to critique it, there has been in practice a noticeable absence of social scientists from practical decision-making on individual natural science-driven projects. Most members of committees set up to review complex technical-scientific issues have a disciplinary background in the natural sciences, especially physics (Roy 2001; Freudenburg and Gramling 2002; Jeronimo 2007). It is as if C. P. Snow's plea for the ascendancy of the 'men of science' over 'literary intellectuals' has finally come to fruition, but with the social sciences now classed as the latter (Snow 1993 [1959]).

Once reduced to being observers rather than prime movers in the technocratic process, social scientists have tried to make a virtue out of necessity and to carve out a distinctive role on and around such scientific committees. Natural resources management committees, studied extensively by Freudenburg and Gramling (2002), serve as a particularly topical example. In their empirical studies, they argue emphatically that the social sciences can make two key specific contributions. The first is to enlighten other scientists on the committee as to the nature, basic principles and findings of social scientific research. They can explain, for example, that even though scientific input may be based on factual and technical matters, any topic in natural resources management also involves values and blind spots. In this view, the public is not necessarily 'irrational', and 'mere perceptions' and blind spots can have very real consequences. What is called 'management of natural resources' is actually the management of human behaviour vis-à-vis those resources.

According to Freudenburg and Gramling, the second contribution social scientists can make is to help colleagues on the committee think about their role in the decision-making process and recognize that there are subtle pressures which may affect the investigation. An example of this is the 'asymmetry of scientific challenge', whereby the results of scientific research are more likely to be challenged when they pose a threat to the groups with an interest in the management of a particular resource. When this occurs, these groups combine to exert systematic pressure on scientific decisions. This is less likely to occur when the discoveries and interpretations are in line with their interests. Social scientists can play a significant role in drawing attention to the extent to which extra-scientific factors can affect scientific assessments – how certain interests may shape the questions, influence the answers obtained and, consequently, affect which facts are regarded as relevant. They may also enlighten the biophysical scientists as to how, at times, their appeals to prudence (when they conclude that 'more research is needed') lead to political actions and resource management decisions which are anything but prudent.

Recommendations such as these are indicative of the modified role social scientists can play in a modern democracy. In a similar vein, social scientists have analysed further public attitudes towards risk and uncertainty and introduced a significant empirical counterweight to the largely deductive 'scientific' approach to decision-making, one which regards the public view as a source of information rather than a remediable error. Appreciation of the public attitude towards an unknowable future, together with refinement of the treatment of risk and uncertainty in policy-making, have opened up a new role for the social sciences to complement judgements for which the natural sciences are too narrow a guide.

Restoring debate by reconceptualising risk

In approaching technical and managerial problems, social and natural sciences frame issues differently. Although they tend to cluster a large range of issues under the category of 'risk', this concept entails contrasting meanings and practices. Whereas natural sciences and engineering seem strongly identified with a probabilistic notion of risk (in which risk is the product of probability of occurrence multiplied with the intensity and scope of potential harm), social sciences interpret 'risk' in terms of uncertainty and contingency. Beck's *Risk Society*, for example, drapes the concept in ambiguity when he designates modern society as a risk society, while using a notion of risk that would be better rendered by the idea of uncertainty. Beck recognizes that there are areas of the unknown, contingency, unpredictability and ignorance when he argues that our society faces "... incalculable threats, which are constantly euphemized and trivialized into calculable risks" (Beck, Giddens and Lash, 1994, p. 182). An undifferentiated mix of the concepts of 'risk', 'uncertainty' and 'ignorance' is consequently absorbed by the generic and institutionalised category of risk.

This mixing might have been avoided if account had been taken of the long-standing discussion of the concepts of risk and uncertainty in economics and political economy. As clarified by Knight (1971[1921]) and Keynes (1921), *risk* refers to situations basically well known, in which probabilistic models can quantify the probabilities of different outcomes, whereas *uncertainty* arises when the probability distributions are not known and are considered the result of incomplete knowledge. More recently, in addition to these two categories and canonical definitions, the notion of uncertainty has been further refined. In cases where we know the possible consequences, but do not know the probabilities, we are faced with a situation of *uncertainty*. A combination of unknown consequences and failure to acknowledge the limitations and compromises of scientific knowledge itself can be classified as *ignorance*. Uncertainties which derive from the existence of contingent social behaviour create a situation of *indeterminacy*, in which it is acknowledged that scientific assessments are the outcome of a particular definition of the problem and that this definition is influenced by social, political and scientific choices (Wynne, 1992, pp. 114–119).

There is a growing awareness that many cases which are treated by natural scientists in terms of probabilities are better viewed as situations of uncertainty, indeterminacy or ignorance, not least because they involve people acting on perceptions which are not risk-based in the scientific sense. The new social studies of science draw attention to these differences between 'ordinary' perceptions and scientific evaluations and argue for the latter to incorporate the former – or at least take account of the way they anticipate, and may further create, unintended consequences to scientific designs. They have helped draw attention to the possibility that public disagreement with scientists' policy advice may arise from valid differences in the way possible impacts are identified and weighted, not just from a misunderstanding of probabilities and expected values. In so doing, they have recalled earlier alternative approaches to decision-making under uncertainty (notably Shackle 1955, 1961), reinforcing them with substantial new empirical evidence.

New approaches in social studies of science also imply that what Beck calls 'risk' is something greater than the index of probability (of potential harm or upside), which the natural sciences put forward. The latter would oversimplify the multidimensional nature of irreversible commitments to complex projects. It would neglect the influence of context and other social, situational and institutional factors, such as the way monitoring mechanisms operate; the equitable distribution of costs and benefits; the credibility of institutions charged with managing 'risks'; or the social stigma attached to a technological facility. The social studies of science accommodate citizens' perceptions, by taking on board their apprehensiveness regarding potential danger to public health and the natural and social environment of a certain technology or industrial facility. It recognises that citizens use 'intuitive risk judgements' and take into account other impacts that fall outside a formal design (Slovic, 2000 [1987], p. 220). The global economic downturn of 2007–9 showed that ordinary savers and borrowers were often more aware of the unsustainable practices of financial institutions than the economic scientists who were advising politicians and sector regulators. This unexpected departure from professional assumptions about financial engineering joined a lengthening list of sometimes devastating departures from official plans in civil, electrical, nuclear, chemical, pharmaceutical, transport and weapons engineering. While experts often blame human error or mistaken judgement for the derailing of their designs (e.g., Perrow 1999), the presence of humans as the operators, overseers and clients of such expertly designed systems require such judgements to be understood and catered to, not left out of account.

Recent social studies of expertly informed public policy-making reassert the view that one cannot conceive of democracy without public debate on scientific issues. There remains a constructive role for 'lay' questioning of the 'scientific' judgements that underlie policies, laws, organizations or implementation procedures which directly or indirectly affect the public. Public debate requires that citizens should take part in this process, on the grounds that decisions should be taken in the name of, and in the interest of, society as a whole. Behind this statement lies the idea that democracy is not just a political system of government, but also a culture, an arena for reconciling divergent interests, an invitation to people to take part in the managing of the *res publica*. The validation of popular risk perceptions, that may differ from those of scientific specialists, appeals to the democratic principle that technological choices should remain subject to a political process – involving public input not just for legitimacy, but because other voices are genuinely needed to make a choice which cannot be determined solely by quantified risks and rewards.

Social studies have thereby highlighted the need for more democratic procedures in technological planning, assessment and implementation. Many natural as well as social scientists argue for the public participation in technical decision-making and for the value of lay knowledge as a way to short-circuit several problems: to enhance public trust in the entities involved in decision-making and control of technology; to attribute greater worth to citizens' opinions on decisions which affect their lives; to encourage democratic practice and to subject institutional interests to public scrutiny (e.g., Otway 1987; Shrader-Frechette 1993; Wynne, 1982, 1992, 1996, 2005; Irwin 1995; Petts 1997; Fischer 2000; Jasanoff 2003). Several proposals have been suggested in order to take account of the various goals and problem definitions of all stakeholders involved, from a 'science court' (so named by the media on the basis of a proposal by Arthur Kantrowitz), 'technology tribunal' (Shrader-Frechette 1985) and 'public space of expertise' (Roqueplo 1997) to 'citizens' conferences' (Fixdal 1997; Boy, Kamel and Roqueplo 2000) and 'extended peer communities' (Funtowicz and Ravetz 1993).

Engaging technocracy in dialogue

The social studies of science and technology evaluation cited above have highlighted the need to think anew and creatively about the relationship between the social sciences and democracy. Once it is recognised that public evaluation of policy outcomes and their risks incorporates valid considerations that fall outside 'scientific' judgement, the dialogical component becomes central to democratic processes. This is especially, but not exclusively, so in the design of technical and organisational systems in which human action is an input and public welfare a targeted output, so that people's views about the system can assist or impede the arrival at the desired outcome and its perception as desired. The notion of democracy implies procedures of open discussion and critique with every member having equal access to the debate. Rather than seeing social research as an auxiliary to the democratic process (as is the case in the technocratic vision), there are arguments in favour of a dialogical notion of social research, one which puts this notion of dialogue at the core of its enterprise.

The contribution of social thought to democratic dialogue has been reinforced by these empirical studies. But social theory makes a further contribution to public debate which, while reinforced by recent empirical findings, is relatively independent of them and in some ways pre-dates them. Many early social theories aspired to replicate the perceived technocratic success of natural scientific theory – building generalised explanatory models from empirical data, thereby showing how social mechanisms worked and could be expertly reengineered for human betterment. The last couple of decades saw a transition from a technocratic approach to the use of theory in a dialogical fashion. Rather than revealing and redirecting the social 'laws of motion' to which agents were subordinate, this dialogical approach draws on social theory to interpret social phenomena, whilst using the confrontation with this empirical material to assess, reconsider and rearticulate the premises of the very same theory. 'Vertical' social theory, whose pioneers viewed themselves as external, neutral observers building a general explanatory framework, was challenged by a more 'horizontal' approach which recognised social scientists as working within society rather than looking down on it, and needing to listen as well as lecture to its subjects (see also Baert 2005, chapter 7).

The dialogical approach can be contrasted with the traditional view of social research, which assumes an asymmetrical relationship between subject and object – between knower and known. In this view, researchers impose their categories on their object of research, and they set out to explain and possibly predict the phenomena under study. They often use analogies with familiar phenomena in order to make sense of the unfamiliar, to absorb it within a known vocabulary and make it comprehensible in existing terms. By doing so, they take away the unfamiliarity of what is being studied – it has been translated into a recognisable language. But in this process, what is being studied has inevitably lost its distinctive voice. Research does not properly engage with the unfamiliar but reinterprets it so that it is no longer able to challenge what is familiar.

The irreconcilability of different deductive models, and their difficulty in accurately forecasting social change or improving its direction, has led various sociological traditions towards a more dialogical notion of social research, which puts the notion of equal, two-way exchange (of questions and answers) at the centre of the relationship between subject and object. In this dialogical view, knowledge is no longer about imposing your own categories onto what is being studied; it is no longer about turning the unfamiliar into something familiar. It is about engaging with otherness – to be open to other voices – so as to make explicit, confront and alter some of our presuppositions and beliefs (Baert 2005, 2009). This type of self-referential research uses the confrontation with difference as an opportunity for communities to become more aware of and potentially challenge various assumptions and ideas which people hitherto took for granted. Research then becomes an ongoing conversation where different voices have, in principle, an equal status and equal chance of being heard.

The dialogical model ties in with a number of philosophical traditions, notably Rorty's neopragmatism, Gadamer's hermeneutics and Levinas's take on phenomenology. Like neopragmatism, the dialogical model rejects the notion of a neutral algorithm that most successful scientific activities supposedly have in common. It also questions the validity of any philosophical search for foundations that would ground cognitive claims. The dialogical model shares with Levinas a concern that most Western thinking fails to properly engage with alterity, and a commitment to making such an engagement possible. Like Gadamer, it promotes the idea that prejudice is a *sine qua non* – not an impediment – for knowledge acquisition but that the latter can also lead to altering the structure of the very same prejudice. Like Rorty and Gadamer, the dialogical model sees *Bildung* or self-edification as a central value underlying any form of inquiry (Baert 2009). It does not deny the validity of natural-scientific approaches or the importance of scientific models and inferences to the design of projects and policy interventions. But in acknowledging the power of subjects to shape the social and economic systems that contain them, it reasserts the necessity of their involvement in public decisions regarding those systems and the continuing role for differences of opinion resolved by political debate.

The democratisation of the relationship between subject and object is also connected to a renewed public role for the social sciences within democratic societies. In the technocratic model, the role of social research is limited to principles of instrumental and means-ends rationality. Social research remains agnostic as to the ultimate values of society. In the dialogical model, social research increases our sociopolitical horizons in that it makes us aware of different sociopolitical scenarios than the ones to which we are used. This increased awareness can then feed into the political process and engender a reflection on the various goals and options that are available.

With the dialogical model, we have new criteria for what constitutes good research. The question is no longer whether the research is adequately predictive or whether it accurately depicts the social realm. Rather, the criterion is whether the research engages properly with the empirical material and whether it enables us to reconsider some deep-seated beliefs – to bring about a *Gestalt*-switch. Research is highly valued if it forces us to reconsider some widespread beliefs, not because those beliefs have been shown conclusively to be false but because the new research increases our imaginative scope and allows us to look at things very differently. What is at stake is the German notion of *Bildung*, which refers to the ability of people to see their previously held views in a much broader light.

This generates a recommendation for writers whom Popperians would regard as speculative and nonscientific. In sociology, Zygmunt Bauman (1991) and Richard Sennett (1998) spring to mind. Their work does not fit neat scientific criteria, but they use their empirical material to challenge contemporary orthodoxies, inviting us to think differently (see also Baert 2007). This is not to say that we do away with empirical evidence and rigorous research. Researchers can only be persuaded to make a *Gestalt*-switch if they are convinced about the arguments presented. But positivists were wrong in postulating that there is a neutral meta-language that would distinguish science from nonscience and that would take the responsibility away from researchers to make the call as to what is proper supported argumentation and what is not.

Technocratic contributions to the democratic process involve experts addressing politicians who are 'accountable' to the public. Observations, interpretations, explanations and advice are conveyed to those charged with taking decisions. The process is dialogical, but involving dialogue within an elite, with equality achieved through differentiation. Specialists in a particular field communicate with policy-forming specialists, one type of expert proposing and another disposing. Popularised versions of the technical material are then presented to those on whose behalf the decisions are taken. This secondary communication may be made by administrative spokespeople, speech-making politicians, journalists or experts writing and speaking in a personal capacity. But unlike the conversation within the elite, this is a top-down and one-way communication, carried out only after the experts have settled on a 'best' way, and rarely conveying to the public all the information made available to the experts. Because of the size and diffuseness of 'the public', it is usually represented by lobbyists who, even where they resist co-option, are necessarily drawn into the elite dialogue and forced into a more hierarchical communication with the mass of non-experts they seek to represent.

In contrast, dialogical contributions to the democratic process involve experts exchanging observations, interpretations, explanations and advice with members of the public. 'Ordinary' people are then trusted, and encouraged, to engage their elected politicians in similar dialogue, rather than letting experts or even interest-group leaders do it on their behalf. The aim is to effect policy change by debate among a wider public whose conclusions are passed upwards to the elite, rather than deliberations among an elite that are later passed down to the public in summary form. For dialogical contributors, popular engagement often precedes academic presentation, with ideas first worked out in 'popular' fashion through wide-circulation articles and debate before being made 'rigorous' into peer-reviewed articles for academic debate. This reverses the sequence of most technocratic contributors, who prefer to establish authority through the technical paper, achieve decision influence through the briefing paper, and write for the newspaper only when the legislation is passed.

The critical potential of dialogical social science

From socialist movements that tried to spark a revolution through radicalising dialogue with the proletariat (and ended up impatiently self-anointing themselves as its vanguard) to the social-justice movements that ineffectively rallied against turn-of-the-century free trade summits, the history of dialogical attempts at rapid social change is not an especially happy one. The consequence of such groups attempting to send dialogically evolved 'grass roots' messages to policy makers is usually incoherence that the experts can ignore, followed by the hasty assembly of a weakly empowered protest leadership that is easily 'co-opted or marginalised' (Klein 2002) by the elite.

If the dialogical power of social activists is weak, that of social thinkers is now commonly viewed as even weaker. With the increased specialisation, use of jargon, quantification and deep archival referencing of modern social science has arisen the plausible suspicion that its dialogue is confined within itself, not usually addressed to the majority outside the academy and not comprehensible to them even when so addressed. "Who, today," asks Zussman (2001), in a 50-year review of David Riesman's (1950) *The Lonely Crowd*, "... reading dreary monographs or struggling through our leading journals, could still imagine sociology supplanting literature?" Literary critic Lionel Trilling had argued in an original review that Riesman's work achieved for sociology the circulation, impact on public opinion and authority to judge public policy and morality which novelists and poets had previously been assigned. Riesman's 1950 book (with Reuel Denny and Nathan Glazer) seemed to confirm this: with a circulation of one million copies through various editions up to 1969, anticipating the critical message of popular fiction writers such as Braine (1957), Yates (1961) and Beattie (1976) and doing so with a comparable social influence when its circulation and intellectual influence were combined. *The Lonely Crowd* became the best-selling sociological text by a considerable margin, giving its social comment as wide an impact as contemporary journalistic commentators. But as Zussman points out, *The Lonely Crowd* rose and fell, no longer attaining the reprint that 'modern classic' fiction from its era was accorded, and now being academically cited less frequently than the contemporary *Social System* (Parsons 1951) or Mills (1951) or later works like Goffman (1959). And no later sociological work has come close to matching its social and political impact.

Whereas Riesman caught the popular imagination with a work that was meticulously written and based on extensive fieldwork, later 'popular' work by the social science professoriate was more apocalyptic than academic. Although McLuhan (1964), Galbraith (1967), Runciman (1983–97), Lumsden and Wilson (1981) and Ferguson (2001) achieved respectable circulation, they reached a much smaller proportion of a much-expanded reading public. The wider propagation of their message was left to popularised accounts or television tie-ins such as McLuhan (1967), Galbraith (1977), Runciman (1998), Wilson (1998) and Ferguson (2008). These reached a wider public, but not as extensively as *The Lonely Crowd*, and lost academic credibility in the process. When ideas from social philosophy and theory entered the public domain, it was – as before Riesman – through fictionalised accounts (e.g., Guare [1990] on social networking, Stoppard [1993] on chaos theory and human interaction, Amis [1991] on the philosophy of time and the human condition) or popularisations that necessarily sacrificed rigour in pursuit of popular reach. The best-selling popular book by a leading academic at the close of the twentieth century was a work of natural science, Hawking (1988); and even in this case, although it is written by the holder of the Isaac Newton's chair, many professorial peers doubt the scientific status of the text and many cynical reviewers doubt the extent to which the several million buyers have read and understood.

However, to chart the influence of social studies solely by their paperback sales is to ignore other key channels by which theory can influence policy by promoting social dialogue. Although the past half-century's expansion of the academy has made it possible for social science to turn inwards, confining dialogue to its own ranks, pressures within it have ensured enough descents and defenestrations from the ivory tower to preserve and magnify its exchange of ideas with the world outside. Contemporary academics do not just write books, although institutional pressure for publication and citation has revived this art. They also now address a growing proportion of the public directly through their degree courses, and get follow-up messages across through consulting for their bosses, advising their politicians, appearing on the traditional media or blogging and podcasting their way into the new ones. Even before the academy began to 'disintermediate' through new online technology, social thinkers who adopted a dialogical approach had found at least four ways to retain an impact beyond the 'dreary monographs' bemoaned by Zussman.

First, dialogical social science identifies problems and problematises existing solutions – a frequently negative enterprise, but one without which technocracy risks missing or misdiagnosing highly significant social problems. The main contribution of academically informed social commentary has often been to crystallise and rationalise inchoate social fears, pinning down a problem so that further enquiry can start to look for solutions, or at least know where they need to be found. In the case of Riesman et al., the achievement was to identify a shift from the internalised values of 'tradition direction' and entrepreneurial spirit of 'inner direction' to the potentially conformist, suggestible 'other direction'. *The Lonely Crowd* thereby offered plausible reasons for a previously unarticulated discontent – felt by bored middle managers, newly rich upwardly mobile workers, their unfulfilled partners and alienated children as much as by the economists and quantitative sociologists unable to determine why unprecedented affluence was not bringing unprecedented harmony.

A century before, similar articulation of previously real but noncodified social stresses had been offered by Marx's theories of alienation, Durkheim's theory of anomie and Tonnies's theory of the shift from community to society. In the half-century after, similar crystallisation and distillation of tacit discontents was achieved by (among others) Galbraith's (1958) *Affluent Society*, Friedmans' (1980) *Free to Choose*, Beck's (1992) *Risk Society* and Etzioni's (2004) *The Common Good*. For the more radical of these authors, explanation was intended to generate more coordinated expression of the discontent arising from social change, so as to take it in a different direction. For the more conservative, it was to draw attention to dangerous undercurrents so that the technocrats could take corrective action. For all, the desired effect was to alert society and its decision-makers to trends emerging from the application of new technological and organisational knowledge which could be averted from bad ends (if not directed towards good) if more fully understood.

Second, the dialogical approach can keep plurality alive by showing that events, trends and 'facts' can always be viewed and interpreted in different ways. From MacGregor (1960) demonstrating to occupational psychologists that Theory Y (of intrinsic work-motivation) could guide company management at least as effectively as Theory X (of necessary sticks and carrots) to Nelson and Winter (1982) showing economists that routine behaviour and Darwinian selection could account for structural change just as well as rational maximisation, analysis that digs deeper into inductive empirical results has repeatedly undermined any one-to-one mapping of events onto explanations. Whereas technocracy seeks to downgrade other scenarios once a particular course of action is identified as most efficient, the dialogical approach tries to keep 'other worlds' open to assessment and possible future implementation.

Third, dialogical enquiry serves as a reminder that 'evidence' is not independent of the ideas of those who gather it, because social enquiry cannot stay disengaged from its subject matter in the way that (arguably) natural-scientific enquiry can. From Rosenthal and Jacobson (1968) on teachers' self-fulfilling perception of children's school performance to Soros (1987) on financial investors' grounding of action in the observed and anticipated actions of others, and Pinker (1994) and Lieberman (1998) on the co-evolution of mind and language, dialogue builds in an acknowledgement of thinking agents' reflexivity – and the inevitable unintended consequences of individual action across space (macro-micro conflicts) and time (path-dependency and emergence).

Finally, even if they stop short (or deny availability) of workable solutions, dialogical researchers can provide new language in which people (and policy makers) describe and assess their situation and its problems. The 'subconscious mind' as a hidden influence on conscious speech and behaviour, the Prisoner's Dilemma situation with its 'zero-sum game' over private strategies, the need for individuals to gain 'ownership' of a solution and 'closure' of a problematic situation, 'moral panics' as artificial concatenation of alarm bells that trigger regressive policy, the necessity for all economic or social exchanges to have a profitable 'business model' – all constitute a naming of a phenomenon by social reasoning which predates and permits subsequent social action. Whereas the articulation of social strains and associated fears requires a persuasive narrative to be threaded through scattered events, the success of naming often arises from terms' diffusion moving ahead of their definition. The keywords allow acknowledgement of the existence and importance of a problem, while enabling different agents to apprehend and act using their own definition.

Concluding remarks

In this chapter, we have argued that social sciences – inspired by the need to prove their professional value, as well as the desire to improve society – developed early on a technocratic model which subsequently foundered on criteria of effectiveness as well as democratic legitimacy; and that a pragmatist-inspired dialogical model has emerged as a coherent and inspiring alternative. We discussed how recent social studies of science, whilst studying the natural sciences, have strengthened the argument for a dialogical approach. Those studies have shown that the social sciences, whilst often excluded from natural science-driven projects, can contribute constructively to debates surrounding those projects, in particular highlighting neglected problems surrounding the implementation of new technologies and the risks they entail. Most importantly, the social studies of science have shown convincingly the importance of taking into account 'lay' perceptions of scientific judgements and of involving citizens within a broader debate about science-driven policies. One of the core arguments of this chapter has been that social research should observe a similar dialogical model whereby researchers no longer impose their categories on what is being studied, but properly engage with otherness, keeping open the possibility of altering some of their core presuppositions and beliefs. This form of research can help broaden people's political scope and thereby assist the democratic process.

As a final observation, it might be worth elaborating on how this dialogical model relates to the gradual decline of grand projects, whether in social theory or in politics. In social sciences, it is often observed – usually with approval – that the era of 'grand theory' is over. Overarching explanations for societies' characteristics across space and development across time (such as those of Marxists, Social Darwinists, Skinnerian behaviourists and Socio-biologists) came under fire for being teleological, historicist and non-verifiable; for misrepresenting or misinterpreting specific events to fit an overgeneralised template; and/or for spanning multiple social disciplines showing a mastery of none. Grand theory went counter to the trends of disciplinary separation and ever deepening specialisation, and those who celebrated its passing viewed social science as having become more 'scientific' as a result. By contrast, the demise of 'grand theory' in politics is often greeted with regret, if not alarm. Personality-based politics is viewed as a step backwards from that based on programmes and general principles. Campaigns fought on the character and media-friendliness of the candidates, rather than what they stand for, are condemned as using stylistic differences to conceal a convergence of substance. In the absence of distinct party principles, clashing figureheads give the illusion of difference while technocrats drive a convergence of policies, giving rise to complaints that "whoever you vote for, the government always gets in." Still more ominously for democracy, politics without distinct programmes can give rise to leaders who are hard to dislodge, because their right to rule becomes inherent rather than objectively assessable by their success in upholding pre-set principles.

Social sciences' critical challenge to earlier grand theories helped to make them a more pluralistic and progressive enterprise, but may inadvertently have encouraged a challenge to grand projects and principles that limits democratic debate. If grand theory invited political rule by lofty demagogues, subsequent granulated theories gave an equally problematic platform to backroom technocrats. Dialogical social science offers a way out, not by scaling the theories up or down, but by fundamentally changing the way they are generated. The challenge for an increasingly educated society, in which knowledge promotes diversity action and reaction that render policy impacts less predictable, is to cease building policies (or theories) over people's heads and instead pay more serious attention to what is in them.

References

Akerlof, G. and J. Yellen, eds (1986) *Efficiency Wage Models of the Labour Market.* Cambridge: Cambridge University Press.

Amis, M. (1991) *Time's Arrow*. London: Vintage.

Ashman, K. and P. Baringer, eds (2001) *After the Science Wars*. London: Routledge.

Baert, P. (2005) *Philosophy of the Social Sciences; Towards Pragmatism*. Cambridge: Polity Press.

Baert, P. (2007) 'Why Study the Social.' In: *Pragmatism and European Social Theory*, ed. P. Baert and B. S. Turner. Oxford: Bardwell Press, pp. 45–68.

Baert, P. (2009) 'A Neo-Pragmatist Agenda for Social Research; Integrating Levinas, Gadamer and Mead'. In: *Pragmatism in International Relations*, ed. H. Bauer and E. Brighi. London: Routledge, pp. 44–62.

Bauman, Z. (1991) *Modernity and the Holocaust*. Cambridge: Polity Press.

Beattie, A. (1976) *Chilly Scenes of Winter*. New York: Vintage.

Beck, U. (1992) *Risk Society: Towards a New Modernity*, trans. Mark Ritter. London: Sage.

Beck, U., A. Giddens and S. Lash (1994) *Reflexive Modernization: Politics, Tradition and Aesthetics in the Modern Social Order*. Cambridge: Polity Press.

Bernstein, R. (1992) *Beyond Objectivism and Relativism: Science, Hermeneutic, and Praxis*. Philadelphia: University of Pennsylvania Press.

Boy, D., D. Donnet Kamel and R. Roqueplo (2000, August–October) 'Un exemple de démocratie participative: la "Conférence de Citoyens" sur les organismes génétiquement modifies.' *Revue française de science politique* 50(4–5), 779–809.

Braine, J. (1957) *Room at the Top*. London: Eyre & Spottiswoode.

Dahl, R. (1998) *On Democracy*. New Haven, CT: Yale University Press.

Elster, J., ed. (1998) *Deliberative Democracy*. Cambridge: Cambridge University Press.

Etzioni, A. (2004) *The Common Good*. Cambridge: Polity Press.

Ferguson, N. (2001) *The Cash Nexus*. London: Allen Lane.

Ferguson, N. (2008) *The Ascent of Money*. London: Allen Lane.

Fischer, F. (2000) *Citizens, Experts, and the Environment: The Politics of Local Knowledge*. Durham, NC: Duke University Press.

Fixdal, J. (1997, December) 'Consensus Conferences as "Extended Peer Groups".' *Science and Public Policy* 24(6), 366–76.

Flyvbjerg, B. (2001) *Making Social Science Matter*. Cambridge: Cambridge University Press.

Freudenburg, W. and R. Gramling (2002, March) 'Scientific Expertise and Natural Resource Decisions: Social Science Participation on Interdisciplinary Scientific Committees.' *Social Science Quarterly* 83(1), 119–36.

Friedman, M. and R. Friedman (1980) *Free to Choose*. New York: Harcourt Brace.

Funtowicz, S. and J. Ravetz (1993, September) 'Science for the Post-Normal Age.' *Futures* 25(7), 739–55.

Galbraith, J. K. (1958) *The Affluent Society*. New York: Houghton Mifflin.

Galbraith, J. K. (1967) *The New Industrial State*. London: Hamish Hamilton.

Galbraith, J. K. (1977) *The Age of Uncertainty*. London: Trafalgar Square.

Goffman. E. (1959) *The Presentation of Self in Everyday Life*. Harmondsworth: Penguin.

Gross, G. and P. Levitt (1997) *The Flight from Science and Reason*. New York: New York Academy of Science.

Guare, J. (1990) *Six Degrees of Separation*. New York: Vintage.

Habermas, J. (1981a) *The Theory of Communicative Action (Volume 1); Reason and the Rationalization of Society*. Boston: Beacon Press.

Habermas, J. (1981b) *The Theory of Communicative Action (Volume 2); Lifeworld and System: A Critique of Functionalist Reason*. Boston: Beacon Press.

Hawking, S. (1988) *A Brief History of Time*. London: Bantam.

Hayek, F. (1945) 'The Use of Knowledge in Society.' *American Economic Review* 35(4), 519–30.

Hayek, F. (1979 [1952]) *The Counter-Revolution of Science*. Indianapolis: Liberty Fund.

Irwin, A. (1995) *Citizen Science: A Study of People, Expertise and Sustainable Development*. London: Routledge.

Jasanoff, S. (2003) 'Breaking the Waves in Science Studies.' *Social Studies of Science* 33(3), 389–400.

Jerónimo, H. (2007) *Scientific Expertise, Uncertainties and Politics: the Protracted Social and Political Conflicts over Hazardous Industrial Waste in Portugal*. Unpublished Ph.D. dissertation in Sociology, University of Cambridge.

Keynes, J. M. (1921) *A Treatise on Probability*. London: Macmillan.

Klein, N. (2002) *Fences and Windows*. London: Picador.

Knight, F. H. (1971 [1921]) *Risk, Uncertainty and Profit*. Chicago: University of Chicago Press.

Kuhn, T. (1970) *The Structure of Scientific Revolutions*. Chicago: University of Chicago Press.

Kukla, A. (1996) 'Does Every Theory Have Empirically Equivalent Rivals?' *Erkenntnis* 44(2), 137–66.

Lash, C., B. Szerszynski and B. Wynne, eds (1992) *Risk, Environment and Modernity; Towards a New Ecology*. London: Sage.

Latour B. and S. Woolgar (1986) *Laboratory Life: The Construction of Scientific Facts*. Princeton, NJ: Princeton University Press.

Lieberman, P. (1998) *Eve Spoke: Human Language and Human Evolution*. London: Norton.

Lumsden, C. and E. O. Wilson (1981) *Genes, Mind and Culture*. Cambridge, MA: Harvard University Press.

MacDonald, K. (1995) *The Sociology of the Professions*. London: Sage

MacGregor, D. (1960) *The Human Side of Enterprise*. London: McGraw-Hill.

McLuhan, M. (1964) *Understanding Media: The Extensions of Man*. London: Sphere.

McLuhan, M. (1967) *The Medium Is the Message*. London: Penguin.

Malkiel, B. (2007) *A Random Walk Down Wall Street*. New York: Norton.

Mills, C. W. (1951) *White Collar: The American Middle Classes*. Oxford: Oxford University Press.

Mommsen, W. (1974) *The Age of Bureaucracy*. Oxford: Blackwell.

Nelson, R. and S. Winter (1982) *An Evolutionary Theory of Economic Change*. Cambridge, MA: Harvard University Press.

O'Neill, O. (2002) *A Question of Trust*. Cambridge: Cambridge University Press.

Otway, H. (1987) 'Experts, Risk Communication, and Democracy.' *Risk Analysis* 7(2), 125–9.

Parsons, T. (1951) *The Social System*. New York: Free Press.

Petts, J. (1997, October) 'The Public-Expert Interface in Local Waste Management Decisions: Expertise, Credibility and Process.' *Public Understanding of Science* 6(4), 359–81.

Perkin, H. (1987) *The Rise of Professional Society*. London: Routledge.

Perkin, H. (1989) *The Rise of Professional Society: England since 1880*. London: Routledge.

Perrow, C. (1999) *Normal Accidents: Living with High-Risk Technologies*. Princeton, NJ: Princeton University Press.

Pinker, S. (1994) *The Language Instinct: How the Mind Creates Language*. New York: William Morrow.

Pinker, S. (1997) *The Language Instinct*.New York: Morrow.

Popper, K. (1971a) *The Open Society and Its Enemies, Volume I; The Spell of Plato.* Princeton, NJ: Princeton University Press.

Popper, K. (1971b) *The Open Society and Its Enemies, Volume II; The High Tide of Prophecy: Hegel, Marx, and the Aftermath.* Princeton, NJ: Princeton University Press.

Popper, K. (1991) *The Poverty of Historicism.* London: Routledge.

Porter, R. (1997) *The Greatest Benefit to Mankind.* London: HarperCollins.

Quine, W. (1975) 'On Empirically Equivalent Systems of the World.' *Erkenntnis 9*, 313–28.

Riesman, D. (1950) (with R. Denny and N. Glazer) *The Lonely Crowd.* New Haven, CT: Yale University Press.

Roqueplo, P. (1997) *Entre savoir et décision; L'expertise scientifique.* Paris: INRA Editions.

Rosenthal R. and L. Jacobson (1968) *Pygmalion in the Classroom.* New York: Holt, Rinehart & Winston.

Roy, A. (2001) *Les experts face au risque: le cas des plantes transgéniques.* Paris: PUF.

Runciman, W. G. (1983, 1989, 1997) *A Treatise on Social Theory.* Cambridge: Cambridge University Press.

Runciman, W. G. (1998) *The Social Animal.* London: HarperCollins.

Schumpeter, J. (1942) *Capitalism, Socialism and Democracy.* New York: Harper & Row.

Sennett, R. (1998) *The Corrosion of Character.* New York: Norton.

Shackle, G. (1955) *Uncertainty in Economics.* Cambridge: Cambridge University Press.

Shackle, G (1961) *Decision, Order and Time in Human Affairs.* Cambridge: Cambridge University Press.

Shipman, M. (1997) *The Limitations of Social Research.* London: Longman.

Shrader-Frechette, K. S. (1985) *Science, Policy, Ethics and Economic Methodology. Some Problems of Technology Assessment and Environmental-impact Analysis.* Dordrecht: D. Reidel Publishing Company.

Shrader-Frechette, K. S. (1993) *Burying Uncertainty. Risk and the Case Against Geological Disposal of Nuclear Waste.* Berkeley: University of California Press.

Sokal, A. (1996a) 'Transgressing the Boundaries: Towards a Transformative Hermeneutics of Quantum Gravity.' *Social Text* 14(1, 2), 217–52.

Sokal, A. (1996b, May–June) 'A Physicist Experiments with Cultural Studies.' *Lingua Franca*, 62–4.

Slovic, P. (2000 [1987]) 'Perception of Risk.' In: *The Perception of Risk*, ed. P. Slovic. London: Earthscan, pp. 220–31.

Snow, C. P. (1993 [1959]) *The Two Cultures.* Cambridge: Canto.

Soros, G. (1987) *The Alchemy of Finance.* New York: Wiley.

Stanford, P. (2006) *Exceeding Our Grasp.* Oxford: Oxford University Press.

Stoppard, T. (1993) *Arcadia.* London: Faber.

Taleb, N. (2001) *Fooled by Randomness.* New York: Norton.

Turner, E., A. Matthews, E. Linardatos, R. Tell and R. Rosenthal (2008, January) 'Selective Publication of Antidepressant Trials and Its Influence on Apparent Efficacy.' *New England Journal of Medicine* 358(3), 252–60.

Weber, M. (1949) *The Methodology of the Social Sciences.* New York: Macmillan.

Weber, M. (1968) *Economy and Society*, ed. G Roth and C. Wittich. New York: Bedminister Press.

Wilson, E. O. (1998) *Consilience; The Unity of Knowledge.* London: Little, Brown.

Wynne, B (1982) *Rationality and Ritual: The Windscale Inquiry and Nuclear Decisions in Britain*, Chalfont St Giles: The British Society for the History of Science.

Wynne, B. (1992, June) 'Uncertainty and Environmental Learning: Reconceiving Science and Policy in the Preventive Paradigm.' *Global Environmental Change: Human and Policy Dimensions* 2(2), 111–27.

Wynne, B. (1996) 'May the Sheep Safely Graze? A Reflexive View of the Expert-Lay Knowledge Divide.' In: *Risk, Environment and Modernity: Towards a New Ecology,* ed. S. Lash, B. Szerszynsky and B. Wynne. London: Sage, pp. 44–83.

Wynne, B. (2005) 'Risk as Globalizing "Democratic" Discourse? Framing Subjects and Citizens.' In: *Science and Citizens: Globalization and the Challenge of Engagement,* ed. M. Leach, I. Scoones and B. Wynne. London: Zed Books, pp. 66–82.

Yates, R. (1961) *Revolutionary Road.* New York: Little, Brown.

Zussman, R. (2001) 'Still Lonely After All These Years.' *Sociological Forum* 16(1), 157–66.

2
Stakeholders or Experts? On the Ambiguous Implications of Public Participation in Science

Stephanie Solomon

Critiques of the social sciences have emphasized the broad and categorical exclusions of certain groups from their discussions and both the political and epistemic ramifications of these exclusions. For example, feminists have pointed out how the exclusion of women's voices have shaped sociological conceptions of labor and rape, psychological conceptions of hysteria, and legal definitions of harassment in ways that both misinterpreted social reality and defined it in ways that benefited men and disenfranchised women. Postcolonial theorists like Edward Said have charged anthropological interpretations of "primitive" peoples with both being complicit with imperialist and colonizing political programs and dividing explanations of non-Western groups into rationalizing denials of difference or interpretations that leave non-Western peoples as completely Other.

A common thread through these prevalent critiques is a combination of *political* critiques of the structure and impact of the social sciences and *epistemic* critiques of the knowledge they claim to produce. In the former case, political reactions to the imperialist, racist, and sexist history of the social sciences are combined with other political motives to democratize the social sciences include remedying the erosion of public trust that has resulted from social research abuses such as the Tuskegee experiments, Willowbrook, and tests of malaria and sexually transmitted diseases on prisoners and soldiers. Another political motive emerges from the argument that while public funds pay for scientific research, public benefits that emerge from research disproportionately benefit a small proportion of the population. This argument has become known as the 10/90 gap as a result of a report published in 1999 by the Global Forum for Health Research that found that only 10% of the annual spending on health research in the United States is allocated toward social and health problems that face 90% of the world's population (Global Forum for Health, 1999).

Epistemically, on the other hand, the social sciences have been charged with employing universal models of human health and behavior that are not locally applicable, presuming an individualist view of human interactions that downplays the roles of communities, and explaining human thought and behavior in terms that are strongly influenced by the worldviews, values, and priorities of particular populations who have a voice in research. Although these charges are often politically motivated as well, they couch their challenges in terms of the subjectivity, bias, irrelevance, and even falsity of scientific claims about the world. Most often, these critiques argue that these epistemic flaws are the result of excluding "lay experts" from scientific discussions.

The idea of democratizing science has emerged as a response to both of these critiques. Although many, including myself, are sympathetic to the political impetus behind democratizing science, this suggestion runs immediately into both conceptual and practical difficulties. While intricately connected, claims that the social sciences are politically unjust and claims that social science findings are epistemically unjustified are distinct. As we shall see, the ambivalence between these two charges manifests analogously in current attempts to theorize and implement democratic responses to these critiques. One prevalent response to both the political and epistemic flaws in the historical trajectory of the social sciences is to demand the inclusion of those excluded groups and their political recognition as stakeholders in naming and explaining the social world in which they live. Another prevalent response is to argue that excluded groups have a form of lay expertise that is epistemically necessary for adequate science. In other words, it is a call for what James Bohman articulates as a "universally distributed expertise" (Bohman, 1999, p. 595).

The tension between these two approaches emerges from the contrast between the egalitarian and inclusive political ideals of democracy on the one hand and the hierarchical and exclusive political organization of expertise on the other. This contrast is ubiquitous in the arguments supporting the democratization of science, both implicitly and explicitly. For example, James Bohman writes,

> Thinking of democracy as a form of cooperative inquiry asks us to shift our thinking about the usefulness of the best forms of social knowledge away from the dominant model of expert authority. Rather than approximating prior standards of "rationality" or "the scientific method," achieving complex and stable forms of knowledge is dependent upon on-going cooperation and thus upon the collective effort to define ideal social conditions and acceptable courses of action. (Bohman, 1999, p. 595)

In a similar vein, Helen Longino argues, "In my view the fate of knowledge rests in how the current monopoly of expertise can be broken and its production redistributed. That seems the challenge of really democratizing science" (Longino 2002, p. 577).

By understanding scientists as having a "monopoly of expertise" that can be "redistributed," or as a "collective effort" based on "on-going cooperation," Longino's and Bohman's statements echo Marx. They portray the democratization of science as putting the means of producing knowledge in the hands of all, just as in a democratic polity the means to "produce" government is ideally in the hands of all citizens.

While both Bohman and Longino articulate the contrast between democracy and scientific methods in terms of expertise, Naomi Scheman defines it differently:

> As we see it, the primary issue is power: who has it and who gets to use it. In a democracy, power is shared: we all have it and have the right to use it. But in traditional research, the thinking seems to be that the research will be more sound, more valid, more "trustworthy" if one entity (the researcher) has more power over another (the "subject"). (Jordan, Gust, and Scheman 2005, p. 44)

These three articulations are intimately connected. As many have pointed out, questions of knowledge and questions of power are intimately linked in both science and society. Those who have the ability to define the terms with which we describe the world, and especially the terms with which we describe our *social* world, have the power to shape, among other things, health and social policy, the justice system, public services, and laws. Thus, whose voices are considered legitimate interlocutors in defining our shared social world has direct political implications for the structural and institutional dynamics within a society. In societies like the United States, where those who are recognized as members of the academic communities are considered the experts on these issues, many question the legitimacy of excluding nonacademic voices from contributing to social scientific discussions and the implications of these exclusions. Arguments to democratize science are based on the premise that entitlement should belong to all citizens with a stake in the social organization of society.

I will argue that a call for democratizing science, if understood as a call to democratize **expertise**, is philosophically incoherent. I will make this argument by first laying out the fundamental principles of democracy through a discussion of the dominant theories of democracy in vogue today. In the next section, I will present three examples – from feminist theory, the sociology of science, and the practical social science approach of community-based research – demonstrating how existing attempts to democratize science often blur the distinction between experts and stakeholders. In the third section, I will examine the epistemic and political ramifications of expertise, contrast the norms of democracies with the norms of expertise, and show how they are incompatible. In the fourth section, I will show how even accepting this argument, there is a way of applying the notion of expertise consistently *and* incorporating nonscientists in the social scientific discussion, not as democratic stakeholders but as experts in their own right.

Ultimately, I will argue how it is possible, and desirable, to 1) democratize (understood as including all stakeholders) many aspects of current science practice and 2) maintain an epistemically coherent (i.e., nondemocratic) notion of expertise in science. I will argue for an understanding of expertise that still responds to the political and epistemic challenges that the social sciences have historically been participating in and that both propagate unjust politics and espouse biased (Western) knowledge as universal. Because it is largely in response to these charges that the move to democratize expertise is so tempting, if I can "fix" the notion of expertise so that it cannot further these ends, then embracing a contradiction of democratizing expertise will not be necessary. In other words, while I will show that in many ways science can and should be democratized, expertise cannot and should not.

The political theories of democracy

Before analyzing the idea of democratizing science, it is important clarify the idea of democracy by itself. In this section, I will identify the general principles that make a theory democratic, and further I will identify a subset of principles that applies to a particular democratic theory, deliberative democracy, that arguably best approximates the social practices of scientists. Thus, I will presume that if deliberative democracy is incompatible with key aspects of scientific organization, I can infer that the other forms will be more so.

According to the Oxford English Dictionary, our modern notion of democracy is "a social state in which all have equal rights, without hereditary or arbitrary differences in rank or privilege." Similarly, the Stanford Encyclopedia of Philosophy defines democracy as "a method of group decision making characterized by a kind of equality among the participants at an essential stage of the collective decision making." What these two bare-bones definitions have in common with most other definitions of democracy are the fundamental principles that democracies are **egalitarian** and **inclusive**, such that all citizens have equal rights to contribute to, and benefit from, the political system.[1]

There are three primary conceptions of democracy in vogue today that go beyond political equality and inclusiveness to articulate further principles that constitute democracies: procedural democracy, constitutional democracy, and deliberative democracy. Procedural democracy is the democratic theory that holds provisional **majority rule** as the fairest way to resolve societal conflicts. This view is called "procedural" because it holds that democracy is legitimate because its voting procedure is fair, not by presuming that its results are right by any independent standard. Procedural democrats allow only two types of rights that limit majority rule: voting equality and right to subsistence, both of which are seen as prerequisites for fair democratic procedures to take place (Gutmann and Thompson 1996, p. 33). This view is roughly espoused by thinkers such as Robert Dahl, Elaine Spitz, Douglas Rae, and David Truman.

Constitutional democracy is a democratic theory that goes beyond procedural democracy to argue that certain rights have priority independently of, and constraining of, majoritarian decisions. These constraining rights are seen as justified in order to produce **fair outcomes** of democratic processes, not merely fair procedures. For example, in procedural democracy, there is no safeguard in place to preserve basic liberties and opportunities in society which may be threatened by majoritarian voting. Exemplars of this approach are John Rawls, Laurence Tribe, and Ronald Dworkin.

Neither procedural nor constitutional democratic principles are plausibly applied to scientific endeavors, and it is doubtful that those who advocate democratizing science are advocating majoritarian voting on scientific questions or assessment of the fairness of scientific results, independent of their validity. As a result, it is to the third major type of democratic theory, deliberative democracy, that I will focus my attention. This theory best approximates the practice of reason-giving and collective argument in science, and thus, is the best contender for a political ideal of science.

The third major theory of democracy is labeled "deliberative democracy." Deliberative democracy, like procedural and constitutional democracies, is also ideally egalitarian and inclusive. It further presupposes "any one of a family of views according to which the public deliberation of free and equal citizens is the core of legitimate decision making and self-government" (Bohman 1998, p. 401). The value of the democracy, and the credibility of its outcome, rests on the fairness of this deliberative process of giving and taking reasons and coming to a collectively acceptable solution, rather than in the process of majoritarian voting as in procedural democracy. The deliberative process is deemed legitimate as long as "through critical argument that is open to the point of view of others, [the citizen] aims to arrive at policy conclusions freely acceptable by all involved" (Young 2001, p. 672).

In their book *Democracy and Disagreement*, Gutmann and Thompson argue for three criteria or fundamental principles that distinguish deliberative democracy from other forms of democracy. First, a deliberative democracy should manifest **reciprocity**, in that "the fair terms of social cooperation include the requirement that citizens or their representatives actually seek to give one another mutually acceptable reasons to justify the laws they adopt" (Gutmann and Thompson 2002, p. 3). The second principle they articulate is **publicity**, which "concerns the agents by whom and to whom the moral reasons are...offered. The agents are typically citizens and public officials who are accountable to one another for their political actions" (Gutmann and Thompson 1996, p. 15). In the case of deliberative democracy, reasons should be made open and public to all citizens.

The third fundamental principle of deliberative democracy, closely interconnected to publicity, is **accountability**. As Immanuel Kant argued, "The obligation to justify policies to those it affects provides the moral basis of the publicity principle" (Gutmann and Thompson 1996, p. 99). The basic premise behind accountability is that, in a deliberative democracy, reasons for political claims and actions must be given to all those who are bound by them, and likewise must be acceptable to that same constituency.

> In a deliberative democracy, representatives are expected to justify their actions in moral terms. In the spirit of reciprocity, they give reasons that can be accepted by all those who are bound by the laws and policies they justify. (Gutmann and Thompson 1996, p. 129)

Threats to the legitimacy of a democracy can be based on challenges to these five basic principles: the inclusivity and equality that should be the mark of any democracy, and the reciprocity, publicity, and accountability of deliberative democracy. The inclusiveness of a democracy is challenged when "citizens" (those who have the right to contribute) do not include all potentially affected parties (stakeholders), such as when women and African Americans were not allowed to vote in the United States. Even if all are in principle included, democracies also can be criticized for failing to give the contributions of citizens equal weight, such as when African Americans were only counted as three-fifths of a person, and thus their population contributed less to the numbers of representatives each state could have in the federal government.

The three further principles of deliberative democracy can be challenged even when all stakeholders are de jure included and formally given equal voice. Iris Marion Young articulates a prevalent criticism of the ideals of deliberative democracies for being unresponsive to the structural imbalances that shape public discourses. "The activist is suspicious of exhortations to deliberate because he believes that in the real world of politics, where structural inequalities influence both procedures and outcomes, democratic processes that appear to conform to norms of deliberation are usually biased toward more powerful agents" (Young 2001, p. 671).

This is an important point because it indicates that the input of stakeholders in political deliberation is not merely a matter of making citizens feel included.[2] Rather, "[a]ctual political deliberation at some time is required to *justify* the law for this society at this time. The reason-giving process is necessary (though not sufficient) for declaring a law to be not only legitimate but also just" (Gutmann and Thompson 2000, p. 164). As such, the justness of the process and the justification of the outcome are intricately connected according to deliberative democratic theories.

Whether or not deliberative democracy is an achievable, or even desirable, ideal, its achievement depends upon its manifestation of some combination of the above values. It ought to be 1) egalitarian, 2) inclusive, 3) reciprocal, 4) publicly disclosed, and 5) accountable to the public. I will focus on deliberative democracy rather than procedural and constitutional democracy in this chapter because its ideal of public justification and criticism is most plausibly applied to scientific practice, where reason-giving among the community and provisional norms are also paramount. The questions at hand are whether 1) it is possible to apply the ideals of democracy to scientific endeavors and 2) if possible, are they desirable to apply to these endeavors. The presumption of those who wish to democratize science must be that the answer to both of these questions is "yes." In the next section, I will articulate three different manifestations of this affirmative answer.

Blurring the distinction

For some, it appears there is no contradiction in applying some combination of the ideals of democracy to science in general, and to scientific expertise in particular. In this section, I will provide three examples in which theorists or practitioners of social science make the argument for democratizing science. First, I will describe the account provided by a feminist epistemologist, Lynn Hankinson Nelson, for blurring the distinction between scientific communities and social communities in general. Second, I will introduce Brian Wynne's famous account of the Cumbrian sheep farmers' interactions with scientists after the Chernobyl nuclear disaster impacted their farming practices. Finally, I will use the example of community-based research practices to exemplify a current practice that aims to democratize science. These three examples will set the stage for my subsequent discussion of the tensions involved in combining ideals of expertise and ideals of democracy in a single idea of democratizing science.

Near the end of her book *Who Knows? From Quine to Feminist Empiricism*, Lynn Hankinson Nelson argues for a broad conception of a scientific community that comes very close to mirroring a notion of a democratic community. "As an 'epistemological community', a community in which knowledge is constructed and shared, and experiences are possible and shaped, our largest community is, in every sense, a science community" (Nelson 1990, p. 314). For Nelson, the way that people out in the world reflectively engage and understand their world is not qualitatively different from the ways that scientists "know" the world. Nelson comes to this conclusion after using Quine and feminist critiques of science to dissolve several traditional boundaries: the boundary between empirical and metaphysical beliefs, the boundary between theory and observation, and ultimately the boundary between science on the one side and common sense, social and political interests, and values on the other.

For Nelson, all our beliefs, whether empirical, metaphysical, political, or otherwise are interlaced and interdependent. Further, a "scientific community" is characterized by its activity of "constructing theories and naming objects, organizing experiences, discovering relationships between our ways of organizing and explaining experience, subjecting beliefs and claims to critical examination, and paying attention to evidence" (Nelson, 1990, p. 315). Insofar as feminist critiques are participants in this type of activity, as are, loosely, most of our commonsensical ways of dealing with the world, then they are all "science" and part of an overarching "epistemic community." They are thus all in principle able to criticize each other, and the metaphysical and theoretical commitments involved are all, more or less, subject to different types of criticism within this overarching community. Nelson's broad view of science can be interpreted as an epistemic democracy where one need only be self-conscious and self-critical about the empirical adequacy of one's cognitive commitments to be a citizen with an epistemic vote. On Nelson's account, the line between deliberative democratic communities and deliberative scientific communities is significantly blurred. Whether scientific or political, all beliefs are justified by a general process of reflection and empirical observation that is practiced to some degree by everyone.

Another famous account of epistemic democracy comes from a sociologist of science. Rather than arguing, like Nelson, that the mode of empirical reasoning found inside and outside the sciences are significantly alike, Brian Wynne uses a case study to argue that when the objects of knowledge are in the public domain, the public should have a right to participate in defining them. Wynne bases this claim on his famous analysis, published in a series of papers in the late 1980s and early 1990s, examining the interactions between Cumbrian sheep farmers and environmental scientists after the Chernobyl disaster. In his initial article "Sheepfarming after Chernobyl: A study in communicating scientific information," Wynne examines the specific case of the impact of the 1986 Chernobyl nuclear accident on a community of sheep farmers living in the Cumbrian region of northern England. This area not only underwent some of the heaviest radiation fallout in Western Europe as a result of the Chernobyl disaster, it also is located close to Sellafield, another nuclear complex that has been criticized for its radioactive discharge, including a nuclear reactor fire in 1957.

Wynne explores the tensions between the scientific expertise and subsequent policies regarding the Cumbrian sheep farmers on the one side and the experiences and interests of the Cumbrian sheep farmers on the other. Due to their extensive experiences with the topographical landscape, the physiological states of their sheep, and previous encounters with nuclear fallout and scientific responses from Sellafield, Wynne argues that the sheep farmers were able to accrue specialized abilities to discriminate and assess many relevant issues when the Chernobyl disaster occurred.

While the policies and regulations that were implemented in the region made perfect sense according to the radiation scientists, the sheep farmers found these policies both overgeneral and often practically inapplicable. For example, scientists ignored or were unaware of local variations in fallout effects between upland and lowland areas as well as variations in farming procedures that had direct impact on the relative contamination of the sheep. Thus, the scientists advocated broad rules on farming and slaughtering that applied equally to all farmers and sheep in the region. The sheep farmers in the area were only too aware of the nuanced circumstances in which the sheep were contaminated, but they had no voice in the policies.

As a result of this misunderstanding by the scientists, regulations resulted in either noncompliance or compliance with severe and avoidable harm to the farmers' livelihood. By collaborating with the Cumbrian sheep farmers on farm management in the region, these problems could have been avoided. Wynne concludes his analysis by arguing

> The allocations of authority and power inherent in routine [scientific] decisionmaking communicate built-in assumptions about what kinds of experience and social groups are worth prior status and which are marginal...Effective communication between technical experts and lay people thus requires them to restructure their regular social relationships. (Wynne 1989, p. 37)

This quote demonstrates that while Wynne challenges which social groups should be included in scientific discussions, he maintains the distinction between "technical experts" and "lay people." He leaves unarticulated the grounds on which to replace the assumptions about who has status to contribute and who is marginal to these types of decisions.

A third example of the democratization of science comes from community-based research. In the article "The Trustworthiness of Research: The Paradigm of Community-Based Research," the authors argue that, "A standard norms approach to research...eliminates valuable sources of expertise readily at hand that can be obtained by including nonacademic stakeholders in all stages of the research" (Jordan, Gust, Scheman 2005, pp. 47–48). This view that stakeholders are experts can be found throughout the literature on community-based research. According to the Agency for Healthcare Research and Quality 2003 Report, community-based participatory research (CBPR) is

> A collaborative research approach that is designed to ensure and establish structures for participation by the communities affected by the issue being studied, representatives of organizations, and researchers in all

aspects of the research process to improve health and well-being through taking action, including social change. (AHRQ report, 2004, p. 3)

The report expands this definition to add three additional qualities of CBPR. First, it involves the "co-learning and reciprocal transfer of expertise." Second, it requires shared decision-making power, and third, it implies the mutual ownership of the processes and products of the research enterprise. The value and determination of collaborating communities is based on a mix of political/democratic values (stakeholders "affected by research," social change, decision-making power, mutual ownership) and epistemic values (expertise). Arguments for stakeholding confer to the public communities, or the public at large, a legitimate voice in the scientific process, just as the political stake of people affected by a democratic governmental structure grants them a legitimate voice in the political process. This "stakeholding" rationale for collaborative research can be found in the definition of CBPR quoted above in its political references to political notions such as "ownership," "representation," and "participation" allocated to those "affected by research."

The arguments for community-based research also appeal to a type of expertise to be found in the public. Those who choose not to jettison all talk of expertise justify stakeholder input because they have some sort of "lay expertise," "insider expertise," "local expertise," or other similar notions, which are either contrasted with scientific expertise or used to replace it. For example, in the definition of CBPR above, there is clear reference to "co-learning and reciprocal transfer of expertise." This mix of rationales has led CBPR practitioners to struggle with internal issues of representation and external issues of recognition. Internally, the complete dissolution of the boundary between the epistemological rights of scientists and the political rights of the public results in the problem of weighing the opinions of the vast number of potential (and legitimate) contributors to decision-making. Because it is unfeasible to include everyone in a CBPR project, which subpopulation is considered to properly represent the stakeholders in a project? Further, CBPR practitioners struggle to be recognized by external funders.

Community-based research is continually challenged by questions raised regarding its validity, reliability, and objectivity for both basic research and evaluation research. The predominance of the scientific method in public health makes it difficult to convince academic colleagues, potential partners, and funders of the value and quality of collaborative research. (Israel et al. 1998, pp. 187–88)

The ambiguous relationship between the epistemic and the political rationales for participatory research is one way to account for the trouble it faces acquiring scientific recognition.

The political theory of science

There are several problems with combining the political structure of scientific expertise with the political structure of democracy in the ways set forth in the last section. As we saw in the above discussion about democracies, deliberative democracies in particular utilize "socially organized deliberation on how best to achieve effective consensual ends" (Bohman 1999, p. 590). This socially organized deliberation is egalitarian, inclusive, reciprocal, public, and accountable. At first blush, there seems to be little in these values that is in tension with the practice of social science. Social epistemologists such as John Hardwig have pointed out that the traditional ideal of science as possibly or even ideally an individual, isolated activity fails to reflect the reality that scientific research is a collective achievement integrally based on collaboration, trust, and socially organized debate among peers (Hardwig 1991). So what makes the "socially organized deliberation" idealized by deliberative democracy different from the "socially organized deliberation" that characterizes the social sciences?

Although values of science overlap these democratic values in some ways, they are not necessary or sufficient conditions for science in the way that they are least necessary for democracy.[3] Sciences in general, and social sciences in particular, rest on the assumption that epistemic authority resides in a subset of people called "experts" who have this authority over the rest of the population, often called "laypersons." This "epistemic" authority is properly understood as a type of political authority, and the question at hand is whether this political authority can be distributed throughout the lay population. In order to answer this question, we must first inquire into the current type of authority that experts wield.

The relationship between experts and laypersons is inherently neither egalitarian nor inclusive, but rather hierarchical and exclusive. Laypersons give experts the authority to define problems, determine the appropriate means to resolve those problems, and implement these means as they see fit. The rational authority accorded to experts presupposes that they have the training and experience to provide specialist information and assess reasons for that information that nonexperts do not. Experts' authority is not justified because they win a majoritarian vote but because nonexperts recognize experts, within a domain, as knowing better than they, and thus voluntarily hand over reasoning and decision-making power and see this deference as legitimate.

This relationship is incapable of reciprocity. Unlike deliberative democracies where all stakeholders can (and ideally must) reciprocally give and receive reasons for decisions, expert decisions are based on reasons that nonexperts are not in a position to *directly* evaluate. Alvin Goldman calls the type of reciprocal evaluation of reasons found in deliberative democracy as "direct argumentative justification." In direct argumentative justification, "a hearer becomes justified in believing an argument's conclusion by becoming justified in believing the argument's premises and their (strong) support relation to the conclusion" (Goldman 2001, p. 94). Goldman contrasts this case with the situation where what he calls "novices" are trying to evaluate the arguments of experts.

> ... however, it is difficult for an expert's argument to produce direct justification in the hearer in the novice/2-expert situation. Precisely because many of these matters are esoteric, N will have a hard time adjudicating between [two different experts'] claims, and will therefore have a hard time becoming justified vis-à-vis either of their conclusions. He will even have a hard time becoming justified in trusting one conclusion *more* than the other. (Goldman 2001, p. 95)

So, in addition to being hierarchical and exclusive, expert political authority is also not reciprocal in the way that deliberative democracies aim to be. Furthermore, if novices/laypersons are not in a position to evaluate expert reasons, then the fundamental democratic principle of public accountability appears to fall away. Even if experts provide their reasons to the public, which they often do in simplified forms, it is not within the background of laypersons to evaluate whether or not these reasons are adequate. Experts therefore are not directly accountable to criticisms of the public within their expert domains in this traditional view of the expert/layperson relationship.

> ... the epistemology of the expert layperson relationship can be focused on the concept of rational deference to epistemic authority. This rational deference lies at the heart of a particular form of power that an expert has and is also the center of the particular form of vulnerability that each of us, as layperson, is in. (Hardwig 1994, p. 3)

The basic dynamics of expertise come in direct conflict with those values that are inherent to democracy in general, such as egalitarian and inclusive relationships, and those that adhere to deliberative democracy in particular, such as reciprocity and public accountability to all citizens. This would leave the goal of democratizing expertise in an incoherent position, because it amounts to undermining expertise altogether. This may be a self-professed goal of many who are responding to the historical criticisms of the social sciences, but they must contend with the well-established intuition in both science and society that some people are in a better position than others to know about certain aspects of our world. Those who wish to democratize expertise must respond to or account for the commonplace intuition that there is something about scientific training above and beyond de facto attributions of authority that legitimate scientific analyses of situations.

Arguments for democratizing expertise in the natural sciences lead to such outlandish implications as society deliberating and voting on whether a person is properly diagnosed with one disease or another, or whether the sun revolves around the earth or the earth revolves around the sun. On the other hand, the question of whether there are legitimate experts in the social sciences is a more difficult question. There are good reasons to believe that there are experts in the domains of human behavior and culture just as there are in the domains of physics and chemistry. To put the descriptions and explanations of social behavior of particular populations up to a general discussion seems just as counterintuitive as doing so in the case of physical phenomena. On the other hand, there is something intuitive to the claim that the type of expertise that characterizes the physical sciences is different from the type that characterizes the social sciences.

I do not believe that those who advocate democratizing science are committed to such counterintuitive conclusions, but this is the ultimate implication of the argument that expertise should be universally distributed. I believe that the real intention behind arguments against the hierarchical and exclusive nature of expertise is not that it is hierarchical and exclusive (i.e., nondemocratic), but rather that the historical demarcation between experts and nonexperts, especially in the social sciences, is both epistemically untenable and political pernicious.

In the real-life contexts where the demarcation of experts takes place, common indicators of expertise are often linked with institutional affiliations, such as attending certain schools or achieving certain degrees. Historically, they have also been linked to social identities such as race, gender, caste, and class. For example, Steven Shapin famously analyses the historical example of gentlemanly veracity in seventeenth-century England. Shapin argues that, in that time, "being a gentleman" was used as an indicator for rational authority in general, while being nongentle and/or female was a negative indicator. The justification given for this connection was that competence and trustworthiness were increased with the economic and social independence that result from social advantage. Social advantage meant that gentlemen had little to gain and much to lose from deception (Shapin 1994).

Miranda Fricker uses Shapin's account to bring credence to her hypothesis (shared by many others):

> [t]here is likely (at least in societies recognizably like ours) to be some social pressure on the norm of credibility to imitate the structures of social power. Where that imitation brings about a mismatch between rational authority and credibility—so that the powerful tend to be given *more* credibility and/or the powerless tend to be wrongly denied credibility—we should acknowledge that there is a phenomenon of *epistemic injustice*. (Fricker 1998, p. 170).

Epistemic injustice, as Fricker defines it, can be the result of the phenomenon of "credibility-overspill" in which indicator properties are seen to convey more credibility than is due as a result of unjustified correlations between general credibility of certain social identities and the rational authority to make competent and trustworthy claims about certain topics. More importantly, this phenomenon can manifest where credibility is withheld from people with certain social identities who *do* have the rational authority to make these types of claims.

> What counts as expertise in many real life cases thus conforms to no transcendent criteria of logic or method, but frequently incorporates popular conceptions (and misconceptions) of relevance and reliability, and all too commonly reflects differences in the social and material positions of disputing parties and decisionmakers. (Jasanoff 2003, p. 159)

In her article "The Politics of Credibility," Karen Jones points out that that "testifiers who belong to 'suspect' social groups and who are bearers of strange tales can thus suffer a double disadvantage. They risk being doubly deauthorized as knowers on account of who they are and what they claim to know" (Jones 2001, p. 158).

Thus, the solution to the historical critiques of expertise is not to democratize it, but to respond to epistemic injustices by locating credibility where it belongs. In the next section, I will argue for a way of demarcating expertise that no longer identifies it with the traditional line between scientists and nonscientists, a line that is complicit with many of the criticisms voiced above. This demarcation will not undermine the reality that nonexperts are not in a direct position to evaluate the reasons of experts (reciprocity), but it will rather argue that 1) in the social sciences, experts are no longer equated with credentialed scientists and 2) nonexperts *are* in a position to determine who the experts in a given question should be (public accountability).

Expertise without credentials

Acknowledging that experts, even in the social sciences, have access to reasons and evidence not available to the rest of us may seem to imply that they are immune to criticism, and the type of authority that they hold is not only not a democratic, but most closely resembles a rational dictatorship. Luckily, the hierarchical situation is not as dire as it would seem, for two reasons. First, there are arguably ways for the public to evaluate experts even if we cannot evaluate their reasons. Second, there are good reasons for broadening those eligible for the designation of "experts" to include many nonscientist communities as well as good reasons to limit the traditional epistemic authority of scientists.

Several theorists argue that there are legitimate ways to rationally evaluate and recognize experts, even from a layperson position. Above, I emphasized the impossibility of "direct" evaluation, but this does not entail that laypeople cannot assess the validity of experts in other ways. Alvin Goldman draws an important distinction between direct and indirect argumentative justification (Goldman 2001, p. 95). Indirect argumentative justification derives from an assessment of a "plausible indicator" of expertise, rather than an inspection of the reasons provided by the expert. For Goldman, these external indicators fall under what he calls "dialectical superiority," which include winning disputes with other experts, comparative quickness and smoothness of responses, as well as credentials and inference to the best explanation. These specific indicators are politically and epistemically problematic for the reasons pointed out by Shapin, Fricker, and Jones in the last section.

John Hardwig suggests another contender for an external indicator for recognizing legitimate experts. "How can B have good reasons to believe that A has good reasons to believe that *p* when B does not himself have evidence of *p*? It's easy – B has good reasons to believe that A has *conducted the inquiry necessary* for believing that *p*" (Hardwig 1985, p. 337 italics mine). Determining precisely what constitutes the appropriate inquiry for particular research questions is outside the scope of this chapter, but, as we shall see, it cannot simply be equated with scientific inquiry.

Traditionally in the sciences, social deliberation and the sharing of reasons take place among experts within the sciences, but those outside of it were considered having no expertise, and thus no relevant voice, at all. In their 2002 discussion paper "The Third Wave of Science Studies: Studies of Expertise and Experience," H. M. Collins and Robert Evans maintain an expert/lay divide, but they argue that this need not mirror the problematic divide between social science experts and the public. They argue for a novel distinction "separating specialist experts, whether certified or not, from non-specialists, whether certified or not" (Collins and Evans 2002, p. 251). Experts, they argue, properly refer to "core-group" scientists, a subset of the scientific community, who are deeply socialized in the experimentation or theorization in a particular area of phenomena. Core groups of scientists should hold a special position among scientists as "experts" on a particular subject, while the assumption that credentialed scientists in general hold an expertise about fields of science distant from their own is based on "mythologies of science" (Collins and Evans 2002, p. 251).[4] They argue that expertise is always, and inherently, local. By emphasizing that experts are determined by a type of specialized training in a localized area, Collins and Evans are able to argue for a new epistemic boundary between scientists in a specialized field and scientists in general:

> The wider scientific community no longer plays any special part in the [expert] decision-making process...the wider scientific community

should be seen as indistinguishable from the citizenry as a whole; the idea that scientists have special authority purely in virtue of their scientific qualifications and training has often been misleading and damaging. Scientists, as scientists, have nothing special to offer toward technical decision-making...where the specialisms are not their own. (Collins and Evans 2002, p. 250)

This argument elucidates Nelson's conception of a broad epistemic community above. The types of evidence and reasons provided for specialized fields are not accessible to everyone, so it is inaccurate to understand our widest community as a community of experts. On the other hand, her argument is supported insofar as the general epistemic orientation of scientists in general does not manifest any authority over general beliefs, and in this way our widest community should be seen as epistemically indistinct from the general scientific community. Thus, while generalized expertise, whether inside or outside of science, does not exist, we can still challenge the epistemic authority held by scientists in general over the epistemic authority of the general population. An important implication of this argument is that the prevalence of "scientific experts" in the media who wax authoritative on many issues outside of their specialty should be especially scrutinized.

Collins and Evans further claim that this understanding of expertise as localized knowledge can be used to challenge the distinction between expert scientists and nonscientist experts on scientific questions. Many scientific endeavors *intrinsically* include hypotheses and theories about how information and actions will play out in public domains. This is paradigmatically the case in the social sciences. Expertise about public domains and expertise about how scientific communities best interact with particular public populations become integrated into scientific questions, methods, and results. Knowledge about public domains, in turn, brings the domain of scientific inquiry beyond the bounds of scientific training and into the bounds of specialized experiences of specific nonscientific populations.

> Once it is recognised that the laboratory is important because it allows scientists a great deal of control, it becomes clear that moving to a real life setting, such as a farm, introduces new complexities that reduce this control and the certainty that it provides (Latour 1983). This is not to say that the science is no longer relevant, but that it can no longer be assumed to be sufficient. Instead, laboratory-based expertise needs to be complimented by the expertise of those with experience in the settings in which it is to be applied. (Evans 2008, p. 284)

Like the core set of scientists, certain groups of people, due to their specific interactions with local environments, have accrued a specialist expertise that has a legitimate voice throughout research about that subject.

Collins and Evans illustrate this point by discussing the aforementioned case of the Cumbrian sheep farmers. They argue that the sheep farmers are misunderstood as merely stakeholders or laypersons, who should have been heeded solely due to the potential effects of scientific policies on their lives and livelihood. The Cumbrian sheep farmers had a specialist expertise in the situation with their sheep, and not merely a stake in the process, although they had that as well. They illustrate this with a thought experiment.

> Imagine that just prior to the Chernobyl explosion a group of London financiers had got together to buy the Cumbrian farms as their private weekend resort, employing the farmers as managers so as to preserve the existing ecology: the financiers, not the farmers, would then be the owners of the sheep, yet all the expertise would remain with the farmers. (Collins and Evans 2002, p. 261)

This thought experiment exhibits one problem with not clearly distinguishing stakeholding from expertise, namely, that stakes are economically and politically transitive while expertise is not. As stakeholders, the sheep farmers were directly affected by the scientific conclusions about acceptable radiation, because this had direct bearing on the welfare and marketability of their sheep, which in turn affected their livelihoods. In spite of this role, Collins and Evans argue that it is not only qua stakeholders that the farmers should have received recognition from the scientific community, but as specialist experts on raising livestock on the Cumbrian fells. Although shifting the ownership of the sheep changes the population of stakeholders, it does not thereby change the population of potential experts about the effects of radiation on sheep. Although the sale of the farms would shift the stakes to the financiers, we would not want to say that the scientists should have then consulted the financiers to learn about the effects of radiation on sheep, although perhaps they should consult them for other things at other stages of the research process.

The case of the Cumbrian sheep farmers discussed above also does not demonstrate that the sheep farmers were "lay scientists," or "lay experts", common phrases used in discourses about democratizing science. Rather "there were not one but two sets of specialists, each with something to contribute. The sheep farmers were not 'lay' anything – they were not people who were not experts – they were experts who were not certified as such" (Collins and Evans 2002, p. 261).

Once we know what we are looking for from experts from the outside (our indirect argumentative justification), we can determine which communities, with which experiences and training, should be recognized as such.

It makes sense to look at expertise as a form of delegated authority, similar to the delegations that legislatures make to administrative agencies.

> By allowing experts to act on their behalf, democratic publics do not give up the right to participate in decisions with a pronounced technical dimension: they only grant to experts a carefully circumscribed power to speak for them on matters requiring specialized judgment. (Jasanoff 2003, p. 158)

Although my argument entails an asymmetry of expertise between different members of a population, we are still not committed to accepting the traditional hierarchies of power and knowledge that have been prevalently criticized in the social sciences. We are, however, committed to some compromise regarding the extent to which knowledge, held by experts, is capable of being democratized. In other words, there is no such thing as "universally distributed expertise," just as there is no such thing as a global expert scientist. Rather, there are specialized core-group scientists with expertise on particular issues. There also is no such thing as an "informed, competent, and ever more emancipated global expert-citizen" (Jasanoff 2003, p. 162).

Concluding lessons

To contrast this hierarchical view of the authority of experts with the egalitarian view of the authority of the general population, David Estlund has coined the imaginative word *epistocracy*. As a political theory, those who have advocated epistocracies can be traced back to Plato's famous Republic ruled by the wise. The idea behind an epistocracy as a political theory is that

> [i]f some political outcomes count as better than others, then surely some citizens are better (if only less bad) than others with regard to their wisdom and good faith in promoting better outcomes. If so, this looks like an important reason to leave the decisions up to them. For purposes of this essay, call them the knowers, or the wise; the form of government in which they rule might be called epistocracy, and the rulers called epistocrats... (Estlund 2003, p. 53)

There are many reasons, contra Plato, to desire that our political systems are democracies and not epistocracies. The social organization in a democracy is organized around the principle that the procedure ought to be fair in that each citizen gets one voice and one vote. Famously, John Stuart Mill argued for a hybrid democratic/epistocratic system of weighted voting, where more votes are given to the better educated. He based this assertion on the somewhat intuitive, if politically incorrect, notion that superior wisdom justifies superior political authority. He also argued that giving more votes to the educated minority is the only way to prevent political decisions from consistently reflecting the unreasoned views of the majority, who by their sheer numbers would not be required to convince anyone of their views.

Estlund's argument against a political epistocracy is that the educated portion of a population may correspond to a population that also has features that make it less worthy to rule, such as race, class, or gender biases, and that there is no systematic way to prevent this. Another, more obvious problem with an epistocracy is that many political decisions regard beliefs about priorities of life, such as what should be taught in schools, whether health care should be universally accessible, or whether there should be laws forbidding gay marriage, abortion, or torturing of prisoners. In the domain of science, questions of which research questions to prioritize and fund and how to disseminate and apply research findings in society also reflect social priorities and values that are indeterminable by specialist expertise. It is unclear how specialized training in a specific area, either within science or outside of it, has any direct bearing on the right one has to determine these issues.

Robert Pierson, in his article "The Epistemic Authority of Expertise," draws a line between the limited and controlled domain in which an expert has expertise and the circumstances in which this knowledge is extrapolated into programs for personal or lay action. These always involve more variables, contexts, and priorities than can be accounted for in the limited domain of expertise. He uses the example of a doctor giving advice to a patient:

> When a doctor diagnoses a patient as having premature ventricular contractions, she is operating within the boundaries of her discipline – she is doing cardiology – and so there is no rational room for lay dissent. However, when the doctor recommends that the patient visit a cardiologist for care of her irregular heartbeat, she is stepping beyond her discipline of cardiology to "advise" that person about treatment. In such cases, she is no longer simply controlling and manipulating her disciplines defining set of variables, but is recommending changes to the layperson's virtually unlimited set of variables. (Pierson 1994, p. 403)

While Pierson is mostly concerned with the contrast between expertise and private life choices, here we are concerned with the contrast between those decisions most appropriately made by the deliberation of a community of experts, whether scientists or nonscientists, and those best made by the deliberation of a community of citizens. Pierson's argument applies to a certain degree here as well because the complex set of variables, contexts, and priorities at stake in a society, while beyond the ken of a closed group of experts, is not necessarily immune to public debate and criticism, that is, deliberative democracy. These issues importantly include many that have traditionally been within the purview of scientific expertise, and those who argue to democratize these aspects of science are making legitimate arguments.

On the other hand, communities of experts are *quintessentially* epistocracies, and democratizing these communities is incoherent. I have argued that laypersons are not in a position to evaluate experts' reasons, because these reasons result from years of theoretical and practical training in observing and analyzing a localized set of phenomena. This same argument applies to "core-group scientists" in a particular domain as well as nonscientists with extensive experience evaluating a particular domain, like the Cumbrian sheep farmers or those with experience as members of particular cultures. Even if the London financiers were put in the exact same geographical location, given the sheep, and given the experience of the radiation fallout, without the years of experience and training that the sheep farmers had on the Cumbrian fells, they would not have known how to interpret and understand the phenomena. Decisions about how to interpret phenomena, whether the manifestations of a disease, the cultural meanings in a particular society, or the effects of radiation on sheep, are not justified by being subject to egalitarian votes among all stakeholders, but rather from conclusions drawn by those with "special aptitudes and skills honed over years of study and practice" (Hardwig 1994, p. 83), whether credentialed scientists or not.

The question that those who advocate democratizing science must answer, then, is this: what aspects of science are properly considered democratic and what aspects are properly considered epistocratic, construed broadly to include nonscientific experts? I have argued that the way to answer this question is to determine, in each stage of exploring a particular research project, whether the questions are those that require specialist training and evaluation of local phenomena – whether atoms, cells, or cultural dynamics – or whether the questions are issues of broad social priorities, social values, and social distribution of scarce resources.

In the former case, it is important for science and society to think about who has relevant experience to provide expertise to a given stage of research. For example, collaborative methods in the social sciences and health have found that methodological issues of recruiting participants, interpretive questions about findings, and even conceptualization issues of determining the domains of relevant and irrelevant phenomena are improved by including nonscientist members of relevant communities as co-researchers in the process. Jean J. Schensul, in her article "Democratizing Science through Social Science Research Partnerships," argues (epistemically) that these research partnerships are vital because

> Most communities, regardless of their socioeconomic status, are able to draw on centuries of constructed, tested local knowledge. We can refer to

this knowledge as "culture." By culture, we mean the accumulated shared repertoire of knowledge, beliefs, and practices that have been found over time to be critical to community survival, reproduction, and protection of future resources. (Schensul 2002, pp. 190–191)

Asking this more specific question about involving relevant expertise could aid CBPR practitioners in their struggles to achieve recognition as scientifically valid, and not just politically more just.

In the latter case, many potential venues exist for garnering public input in a democratic process of assessing research priorities, proper research protections, and proper modes of dissemination.

Social scientists have responded to the triumph of technique and attendant electoral decline by advocating and designing increasingly sophisticated techniques of their own to re-establish a role for non-experts in scientific, environmental, and technological decision making. These include focus groups, citizen juries, community advisory boards, consensus conferences, and participatory integrated assessment. (Jasanoff 2003, p. 164)

I am not arguing that these two sets of questions do not overlap or that possibly both bear on the same stages of research. But I do believe that sensible lines can be drawn in the discussion of democratizing science by asking whether the point is to incorporate relevant nonscientist experts into the expert discussion or to incorporate stakeholders into a discussion where scientists and nonscientists are properly understood to be on even ground. This more specific question, rather than blurring the distinction between expert contributions and stakeholder contributions to the scientific enterprise, makes the distinction, while still addressing both the political and empirical criticisms of the exclusionary history of science.

Notes

1. Even in representative democracies, the ideal that each representative is answerable to his or her constituency maintains these egalitarian and inclusive ideals.
2. Some argue that in science, for instance, the push toward public engagement is aimed more toward making the public feel more familiar with science, rather than to facilitate any actual collaboration.
3. There is a strong debate in the deliberative democracy literature about whether more substantive principles are necessary in addition to the procedural ones that dictate the type of deliberation in place. For one argument that procedural principles are not sufficient, see Gutmann and Thompson 2000.
4. Many (including Wynne) take issue with Collins/Evans presumption that the core group is in some way nonnegotiable itself. I do not believe that this is implied by their argument, but I take the important (and relatively noncontroversial point) to

be that the core group is legitimated by a set of experiences and socialized training with a localized set of phenomena.

References

AHRQ Report: Viswanathan, Meera, et al. (2004) 'Community-based Participatory Research: Assessing the Evidence,' No. 99. AHRQ Publication No. 04-E022-2.

Bohman, J. (1998) 'Survey Article: The Coming of Age of Deliberative Democracy.' *Journal of Political Philosophy* 6(4), 400–25.

Bohman, J. (1999) 'Democracy as Inquiry, Inquiry as Democratic: Pragmatism, Social Science, and the Cognitive Division of Labor.' *American Journal of Political Science* 43(2), 590–607.

Collins, H. M. and R. Evans (2002) 'The Third Wave of Science Studies: Studies of Expertise and Experience.' *Social Studies of Science* 32(2), 235–96.

Estlund, D. (2003) 'Why Not Epistocracy?' In: *Desire, Identity and Existence: Essays in Honor of T. M. Penner.* Kelowna BC: Academic Printing and Publishing, pp. 53–69.

Evans, R. (2008) 'The Sociology of Expertise: The Distribution of Social Fluency.' *Sociology Compass* 2(1), 281–98.

Fricker, M. (1998) 'Rational Authority and Social Power: Towards a Truly Social Epistemology.' *Proceedings from the Aristotelian Society* 98(2), 159–77.

Global Forum for Health Web site (1999) http://www.globalforumhealth.org/Media-Publications/Publications/10-90-Report-on-Health-Research-1999. Accessed 15 January 2009.

Goldman, A. (2001) 'Experts: Which Ones Should You Trust?' *Philosophy and Phenomenological Research* 63(1), 85–110.

Gutmann, A. and D. Thompson (1996) *Democracy and Disagreement.* Cambridge, MA: The Belknap Press of Harvard University Press.

Gutmann, A. and D. Thompson (2000) 'Why Deliberative Democracy is Different.' *Social Philosophy and Policy* 17, 161–80.

Gutmann, A. and D. Thompson (2002) 'Deliberative Democracy Beyond Process.' *Journal of Political Philosophy* 10(2), 153–74.

Hardwig, J. (1985) 'Epistemic Dependence.' *The Journal of Philosophy* 82(7), 335–49.

Hardwig, J. (1991) 'The Role of Trust in Knowledge.' *The Journal of Philosophy* 88(12), 693–708.

Hardwig, J. (1994) 'Toward an Ethics of Expertise.' In: *Professional Ethics and Social Responsibility,* ed. D. Wueste. Lanham, MD: Rowman & Littlefield, pp. 83–102.

Israel, B. A., A. Schulz, E. Parker, and A. Becker (1998) 'Review of Community-Based Research: Assessing Partnership Approaches to Improve Public Health.' *Annual Review of Public Health* 19, 173–202.

Jasanoff, S. (2003) '(No?) Accounting for Expertise.' *Science and Public Policy* 30(3), 157–66.

Jones, K. (2001) 'The Politics of Credibility.' In: *A Mind of One's Own,* 2nd ed, ed. L. Antony and C. Witt. Boulder, CO: Westview Press.

Jordan, C., S. Gust, and N. Scheman (2005) 'The Trustworthiness of Research: The Paradigm of Community-based Research.' *Journal Metropolitan Universities* 16(1), 37–57.

Longino, H. (2002) 'Reply to Philip Kitcher.' *Philosophy of Science* 69(4), 573–8.

Nelson, L. H. (1990) *Who Knows: From Quine to a Feminist Empiricism.* Philadelphia: Temple University Press.

Pierson, R. (1994) 'The Epistemic Authority of Expertise.' *PSA: Proceedings of the Biennial Meeting of the Philosophy of Science Association* 1, 398–405.

Schensul, J. (2002) 'Democratizing Science through Social Science Research Partnerships.' *Bulletin of Science, Technology & Society* 22(3), 190–202.

Shapin, S. (1994) *A Social History of Truth: Civility and Science in Seventeenth-Century England.* Chicago: University of Chicago Press.

Wynne, B. (1989) 'Sheepfarming after Chernobyl: A Case Study in Communicating Scientific Information.' *Environment* 32(2), 11–15; 33–9.

Young, I. M. (2001) 'Activist Challenges to Deliberative Democracy.' *Political Theory,* 29(5), 670–90.

3
Scientific Knowledge: A Stakeholder Theory

Kristina Rolin

Introduction

A stakeholder theory of scientific knowledge addresses the question of whether outsiders to particular scientific communities are allowed to play a role in epistemic justification, and if they are, what this role might be. By stakeholders of scientific knowledge I mean outsiders to particular scientific communities who have an interest in the knowledge produced in these communities.[1] Those who fund scientific research or are in a position to make funding decisions on the behalf of some larger social body, either public or private, are stakeholders for the obvious reasons that it is their or their employer's money that is used in science. Among stakeholders are also laypersons who have an interest in scientific knowledge because scientists study them or because their lives are influenced by the applications of scientific research. Some laypersons, such as AIDS activists and environmental activists, have tried to interfere in scientific debates (see e.g., Epstein 1996; Jasanoff 2004 and 2005; Leach, Scoones, and Wynne 2005).[2] Yet another group of stakeholders consists of scientists who are outsiders to a specialty and nevertheless have an interest in the specialty because of its relevance for their research. Communication and collaboration across the boundaries of specialties have sometimes been a crucial factor in the epistemic success of scientific inquiry (see e.g., Thagard 1999). Professionals such as physicians, therapists, lawyers, engineers, and to some extent corporate managers, are also stakeholders because their status as professionals is dependent on scientific knowledge. Insofar as corporations hope to make use of scientific research in their pursuit of profit or corporate social responsibility, they are stakeholders of scientific knowledge (see e.g., Mirowski 2004). The stakeholder interest in scientific knowledge gives rise to several *normative* questions: Do scientists have a duty to engage in scientific debates with stakeholders under certain conditions? Are scientists entitled to refuse to participate in such debates? What are the epistemic responsibilities of stakeholders and scientists in such debates?

Harry Collins and Robert Evans suggest that the so-called "third wave" in social studies of science should address the normative question of how far citizen participation in scientific debates can legitimately be extended (2002, p. 237). Philosophers of science have been slow to uptake this challenge. Traditionally, philosophers of science acknowledge that outsiders to particular scientific communities can legitimately have a say in issues of significance in science (Kitcher 2001; see also Hempel 1965). In the traditional view, outsiders are allowed to play a role in decisions about what research topics are funded and for what ends scientific knowledge is used. But outsiders have been assumed to lack authority in issues of epistemic justification. The emergence of social epistemology in philosophy of science has not challenged the tradition in this respect. Most social epistemologists focus on social relations within scientific communities, assuming as Thomas Kuhn does that "the members of a given scientific community provide the only audience and the only judges of that community's work" (1996, p. 209).[3] Philosophers who discuss stakeholder relations typically address two questions. One question is how nonexperts can evaluate the trustworthiness of scientific experts (Goldman 2002, pp. 139–163). Another question is what policy issues should be delegated to scientific experts and what should be kept within the sphere of public deliberation. The latter question is motivated by the concern that scientific experts undermine a liberal democratic order insofar as they make decisions on policy issues that properly belong to the sphere of public deliberation (Turner 2006, p. 159).

The question of whether stakeholders are allowed to participate in scientific debates has not been addressed in any systematic way in philosophy of science. My explanation for this is that the stakeholder question is caught in the following dilemma. If the so-called outside criticism to scientific knowledge is to be *effective* criticism, then it is not *outside* criticism at all; it is more properly called *inside* criticism. I call this dilemma *the paradox of outside criticism*. Indeed, given the paradox of outside criticism, the stakeholder question appears to be a nonissue in philosophy of science because it is subsumed under the general question of what the standards of argumentation ought to be in the sciences.

In this chapter, I argue that the paradox of outside criticism is false. The stakeholder question cannot be subsumed under a general theory of the standards of argumentation in the sciences. It is a philosophical problem in its own right. As a representative of a stakeholder theory of scientific knowledge I discuss Helen Longino's social epistemology in *The Fate of Knowledge* (2002). I focus on her argument for the normative claim that scientific communities should be open to outside criticism. This claim makes Longino's social epistemology stand out as one kind of a stakeholder theory of scientific knowledge.

In the first section, I explain Longino's social epistemology. In the second section, I introduce a contextualist theory of epistemic justification and argue that it provides an epistemic justification for the claim that scientific communities should be open to outside criticism. In the third section, I explain why the paradox of outside criticism is false. Finally, I defend Longino's social epistemology against the objection that it lacks naturalistic justification (Solomon and Richardson 2005).

Helen Longino's social epistemology

Longino's social epistemology makes room for the social and economic context of scientific knowledge in two ways: (1) by claiming that epistemic justification is relative to a context of background assumptions and (2) by claiming that outsiders to particular scientific communities can legitimately play a role in epistemic justification. For clarity, I call the first claim the thesis of contextual evidence and the second claim the thesis of outsider participation. In this section, I explain Longino's arguments for these two claims.

Longino argues that we should accept the thesis of contextual evidence because background assumptions are required to establish the relevance of empirical evidence to a hypothesis or a theory (1990, p. 43). In most cases, an observed state of affairs in itself does not tell for what hypothesis or theory it can be taken as evidence. An observed state of affairs can be taken as evidence for quite different and even conflicting hypotheses given appropriately conflicting background assumptions. Also, different aspects of an observed state of affairs can serve as evidence for different and even conflicting hypotheses given appropriately conflicting background assumptions (1990, pp. 40–43). Therefore, Longino argues, "A state of affairs will only be taken to be evidence that something else is the case in light of some background belief or assumption asserting a connection between the two" (1990, p. 44).

Longino acknowledges that the thesis of contextual evidence seems to generate a problem of epistemic relativism (1990, p. 61). As she explains, "By relativizing what counts as evidence to background beliefs or assumptions, hypothesis acceptance on the basis of evidence is also thus relativized" (1990, p. 61). And she admits that "without some absolute or nonarbitrary means of determining acceptable or correct background assumptions there seems no way to block the influence of subjective preferences" (1990, p. 61). As an antidote to epistemic relativism, Longino introduces a social account of objectivity. In her view, objectivity is a characteristic of a community's practice of science rather than of an individual's. One function of a community's practice of science is to critically examine the background assumptions in light of which observed states of affairs become evidence. According to Longino, a critical examination of background assumptions is required to ensure that scientific knowledge is independent of any individual's subjective preferences (1990, pp. 73–74).

In Longino's view, objectivity is a matter of degree. Those communities that fulfill certain norms to a high degree are more objective than those communities that fulfill them to a lesser degree (or do not fulfill them at all) (Longino 1990, p. 76). The purpose of these norms is to ensure that criticism will be effective in bringing about transformation in scientific debates. Longino proposes four such norms (1990, pp. 76–81; see also 2002, pp. 129–131): (1) There must be publicly recognized forums for the criticism of evidence, of methods, and of assumptions and reasoning. (2) There must be uptake of criticism. (3) There must be publicly recognized standards by which theories, hypotheses, and observational practices are evaluated and by appeal to which criticism is made relevant to the goals of the inquiring community. (4) Communities must be characterized by the equality of intellectual authority.

The thesis of outsider participation, the claim that outsiders to particular scientific communities can legitimately play a role in epistemic justification, is implicit in Longino's discussion of the fourth norm. By the equality of intellectual authority, Longino refers to three different dimensions in the social organization of science. First, she discusses equality in access to scientific education and profession, especially as it concerns minorities and non-minority women (1990, p. 78; 2002, p. 132). Second, she discusses equality in relations among those who have made it into the scientific profession (1990, p. 78; 2002, p. 131). Third, she discusses the relation between scientists and nonscientists (2002, pp. 133–134). Whereas the first two dimensions are already present in Longino's *Science as Social Knowledge* (1990), the third dimension emerges in *The Fate of Knowledge* (2002).

In *The Fate of Knowledge* (2002), Longino stresses that, by intellectual authority, she does not mean the same thing as cognitive authority. According to Longino, cognitive authority is domain-specific; it has to do with the amount of knowledge one has of a subject matter (2002, p. 133, note 19). Intellectual authority, by contrast, is less a matter of having knowledge than of "having cognitive or intellectual skills of observation, synthesis, or analysis, which enable one to make cogent comments about matters concerning which one knows less than another" (2002, p. 133, note 19). Longino suggests that a general capacity to argue and evaluate arguments is sufficient to give a person some intellectual authority even in those domains where she does not have cognitive authority.

Also, Longino claims that the demand for the equality of intellectual authority raises complex questions concerning community membership (2002, p. 133). "It requires both that scientific communities be inclusive of relevant subgroups within the society supporting those communities and that communities attend to criticism originating from 'outsiders'" (2002, pp. 133–134). Thus, Longino holds the view that the social practice of epistemic justification should be open to "outsiders' points of view," including not only scientists from other disciplines and specialties but also some nonscientists (2002, p. 134).

However, the first three norms in Longino's social account of objectivity set certain conditions for outsider participation in scientific debates. The first norm requires that outsiders respect the public nature of scientific debates.

The second norm requires that outsiders be responsive to criticism coming from the members of a scientific community. The third norm requires that outside criticism appeal to at least some of the standards of argumentation that are recognized in the target community. Longino emphasizes that the third norm is meant to protect scientific debates from what she calls a "cacophony" of more and less educated voices (2002, p. 133). As she explains, the demand for equality does not mean that every individual is granted equal authority on every subject matter (2002, pp. 132–133). Insofar as outsiders have a right to present criticism in a scientific debate, this right entails duties defined in the first three norms in Longino's social account of objectivity. These limitations to outsider participation enable scientific communities to balance between two extreme situations: a community sealed from outside criticism, on the one hand, and an untended garden where hundreds of flowers are allowed to bloom, on the other hand.[4]

Indeed, Longino suggests that creationists do not qualify as responsible stakeholders of scientific knowledge because they fail to meet all other norms in her social account of objectivity except the first one (2002, pp. 158–159). They fail to meet the second and the third norm because they do not respond to the criticism of the theory of intelligent design. Even though they claim that their religious beliefs are relevant for scientific theories, they are not willing to subject these beliefs to critical scrutiny. They also fail to meet the fourth norm because they do not treat nonbelievers as fully equals to their community of believers. Thus, the four norms in Longino's social account of objectivity are meant to be mutually binding. Not only scientific communities but also other communities are expected to respect them insofar as they make claims to scientific knowledge.

Why does Longino think that we should accept the thesis of outsider participation, the claim that outsiders to particular scientific communities can legitimately play a role in epistemic justification? Longino presents two arguments for the fourth norm in her social account of objectivity, the claim that communities must be characterized by the equality of intellectual authority. One argument claims that the equality of intellectual authority is needed to disqualify those communities where certain perspectives come to dominate because of the political, social, or economic power of their adherents (Longino 1990, p. 78; 2002, p. 131). Another argument claims that the equality of intellectual authority is desirable because it promotes a diversity of perspectives and a diversity of perspectives is necessary for a critical dialogue (Longino 2002, p. 131; see also Mill 1962).[5] The second argument differs from the first one in that it emphasizes the need for active measures to promote a diversity of perspectives in scientific debates. A lack of suppression alone does not guarantee a diversity of perspectives if a scientific community does not include scientists with different perspectives on the subject matter of inquiry. As Longino explains, "Not only must potentially dissenting voices not be discounted; they must be cultivated" (2002, p. 132). She is concerned with equality in access to scientific education and profession precisely because she believes that equality in this sense increases the likelihood that a community of scientists includes a diversity of perspectives. Similarly, Longino is concerned with the relation of a scientific community to outsiders because she believes that outsiders' perspectives are sometimes needed to bring cognitive diversity into scientific debates.[6]

To summarize, Longino argues that we should accept the thesis of outsider participation because it promotes a diversity of perspectives. In her view, a diversity of perspectives is necessary to ensure that background assumptions are critically examined before they are accepted. A critical examination of background assumptions plays a crucial role in the epistemic justification of scientific knowledge because background assumptions are needed to establish the relevance of empirical data to theories and hypotheses.

A contextualist theory of epistemic justification

In this section, I introduce a contextualist theory of epistemic justification, and I argue that it is an apt theory of epistemic justification in science because it enables one to avoid both the problem of relativism in epistemic justification as well as a traditional problem associated with a foundationalist theory of epistemic justification. The purpose of this discussion is to prepare ground for the claim that a contextualist theory of epistemic justification provides an epistemic justification for the thesis of outsider participation. I argue that an epistemically responsible scientist has a duty to respond to outside criticism in certain circumstances. My argument gives a more compelling reason to accept the thesis of outsider participation than Longino's argument. The epistemic importance of outside criticism is not contingent on whether it promotes a diversity of perspectives, as Longino suggests. Responsiveness to outside criticism is an epistemic obligation that scientists have in certain circumstances.

As Michael Williams defines it, a contextualist theory of epistemic justification includes the view that epistemic justification takes place in a context of assumptions, some of which are justified in virtue of functioning as default entitlements and others in other ways (2001, pp. 226–227). Assumptions that function as default entitlements are not plain assumptions; instead, they are assumptions adopted with a defense commitment. A defense commitment means that one accepts a duty to defend or revise one's belief if it is challenged with an appropriate argument. As Williams explains, default entitlements can be articulated and challenged but only by a recontextualization of inquiry that involves assumptions of its own (2001, p. 227). So, even if one can challenge an assumption that functions as a default entitlement, it does not follow that one can transcend the contextual nature of epistemic justification. When one accepts such a challenge, one's inquiry is moved into another context of assumptions; it does not become less contextual.

Williams's contextualism and Longino's social epistemology share the view that epistemic justification is relative to a context of background assumptions (the thesis of contextual evidence). Yet contextualism adds two important qualifications to Longino's social epistemology: one concerns the epistemic status of empirical evidence and another concerns the epistemic status of background assumptions. Longino calls her theory of scientific knowledge "contextual empiricism" because it treats empirical evidence as "the basis of knowledge claims in the sciences" (1990, p. 219). Also, contextualism grants a special status to empirical evidence. Observation reports are paradigmatic cases of beliefs that function as default entitlements in many contexts of inquiry. Scientists are justified in believing in them as long as no one has provided reasons to suspect that they are false or that they have been produced in an unreliable way. In contextualism, empirical evidence is thought to be the basis of knowledge claims in the sense that it functions as a default entitlement in many contexts of inquiry, not in the sense that it is intrinsically justified in virtue of being empirical.

Another qualification added by Williams's contextualism concerns the epistemic status of background assumptions. Contextualism emphasizes that at least some of the assumptions that constitute a context of inquiry are justified in virtue of functioning as default entitlements. They are not adopted dogmatically, even though they are not under scrutiny in the context of inquiry that they constitute. A contextualist understanding of background assumptions is implicit in Longino's social epistemology. For example, in *Science as Social Knowledge,* Longino suggests that background assumptions be treated as default entitlements when she writes that "as long as background beliefs can be articulated and subjected to criticism from the scientific community, they can be defended, modified, or abandoned in response to such criticism" (1990, pp. 73–74). In *The Fate of Knowledge*, Longino suggests that standards of argumentation be treated as default entitlements when she writes that they "are not a static set but may themselves be criticized and transformed, in reference to other standards, goals, or values held temporarily constant" (2002, p. 131).

I argue that a contextualist theory of epistemic justification is an apt theory of epistemic justification in science for two reasons. One reason is that it enables one to avoid the problem of relativism with respect to epistemic justification. Another reason is that it enables one to avoid a traditional problem associated with foundationalism as a theory of epistemic justification. Let me explain these two virtues of contextualism.

By relativism I mean the view that epistemic justification is relative to a context of background assumptions that are treated as plain assumptions (see also Crasnow 2003; Rolin, forthcoming). Contextualism differs from relativism in that it treats background assumptions as default entitlements and not as plain assumptions. To treat an assumption as a default entitlement means that it is adopted with a defense commitment, that is, a commitment to defend it in case it is challenged with contrary evidence or other arguments. As Williams explains, contextualism implies a default and challenge model of epistemic justification (2001, p. 25). In a default and challenge model, an entitlement to one's belief is the default position, but one has a duty to defend or revise it as soon as it is challenged with appropriate arguments (Williams 2001, p. 149). So, whereas both contextualism and relativism include the view that epistemic justification is relative to some context, only relativism includes the view that contexts are "frameworks of ultimate commitments" (Williams 2001, pp. 224–225). In contextualism, no context includes ultimate commitments which are beyond criticism (Williams 2001, pp. 226–227).

By foundationalism I mean the view that a belief is justified if and only if it is either itself a basic belief or inferentially connected (in some appropriate way) to other justified beliefs (see also Williams 2001, p. 164). Foundationalism includes two further assumptions: (1) there are beliefs that are in some sense justifiably held without resting on further evidence, and (2) these beliefs are basic beliefs, that is, beliefs that in virtue of their content are justifiably held without resting on further evidence. A traditional problem in foundationalism is to give an account of why some beliefs are basic beliefs. Contextualism does not need to address this problem because it rejects the second assumption, the claim that there are basic beliefs. However, contextualism is consistent with the first assumption, the claim that there are beliefs that are justified without resting on further evidence. In contextualism, beliefs that are justified without resting on further evidence have this epistemic status in virtue of functioning as default entitlements, not in virtue of their content alone.

Contextualism differs from foundationalism in another important respect. Whereas in foundationalism epistemic justification is a matter of having adequate grounds for a belief, in contextualism epistemic justification is a matter of epistemic responsibility (Williams 2001, pp. 22–25). In order to understand why epistemic responsibility is not the same thing as having adequate grounds, it is necessary to make a distinction between two different conceptions of epistemic justification. When we ask for epistemic justification, we can ask under what conditions a person is justified in believing a particular proposition, or we can ask under what conditions the proposition she believes is justified. Whereas the first question inquires whether a belief has been responsibly formed or is responsibly held, the second question inquires whether a belief has adequate grounds. Traditionally, it has been assumed that one is epistemically responsible in believing that p *only if* one's belief that p is based on adequate grounds (Williams 2001, p. 24). In contextualism, epistemic responsibility is an independent conception of epistemic justification in the sense that a person can be epistemically responsible in believing that p *even if* she does not have adequate grounds for p (Williams 2001, p. 25).

Finally, I am in a position to argue that a contextualist theory of epistemic justification gives an epistemic justification to the thesis of outsider participation, that is, the claim that outsiders to particular scientific communities can legitimately play a role in epistemic justification. In contextualism, a scientist is epistemically responsible in her belief if she presents sufficient evidence in its support or if she adopts a defense commitment with respect to it (Williams 2001, p. 25). In either case, her being epistemically responsible is dependent on the dialogical features of her situation. What counts as sufficient evidence depends on what challenges others actually pose to her argument. Similarly, whether her belief is justified in virtue of functioning as a default entitlement depends on whether the potential challengers are able to raise appropriate objections to it. It follows that an epistemically responsible scientist has a duty to respond to outside criticism insofar as it includes an appropriate challenge to her views. A challenge to a claim is appropriate if it provides a reason to suspect that the claim is false or that the claim has been acquired in an unreliable way. Thus, outsiders to particular scientific communities can legitimately play a role in epistemic justification insofar as they are capable of challenging a scientist's view. Their right to intervene in scientific debates is not unconditional. They have to carry the burden of proof.

To summarize, I have argued that we should accept the thesis of outsider participation because an epistemically responsible scientist has a duty to respond to outside criticism insofar as outsiders actually pose a challenge to her views. In any scientific community, a scientist may be tempted either to ignore or reject out of hand some challenges simply because they come from outside her specialty. However, such gestures would be violations of the normative ideal of epistemic responsibility. Epistemic responsibility requires that a scientist respond to outside criticism if she is challenged in an appropriate way. Epistemic responsibility requires also that the challenger carries the burden of proof and gives a reason to suspect that something is wrong in a scientist's view. Thus, the thesis of outsider participation is supported by a contextualist theory of epistemic justification. This result raises the additional concern whether a challenge coming from outside a specialty is properly speaking an instance of "outside criticism" rather than an instance of "inside criticism." This is the topic of the next section.

Why is the paradox of outside criticism false?

Longino's social epistemology is potentially undermined by what I have called the paradox of outside criticism. The paradox of outside criticism is the claim that, if the so-called outside criticism is to be effective criticism, then it is more appropriately called inside criticism than outside criticism. The paradox of outside criticism is a challenge to the thesis of outsider participation, the claim that outsiders to particular scientific communities can legitimately participate in epistemic justification, because it implies that the thesis of outsider participation is a self-defeating principle. It is self-defeating because as soon as the so-called outsiders participate in epistemic justification, they are no longer outsiders. If the thesis of outsider participation is self-defeating, then openness to outside criticism cannot function as a criterion distinguishing more objective communities from less objective ones. The equality of intellectual authority can function as an epistemic ideal in scientific communities merely in a limited sense, as equality among the members of scientific communities.

In this section, I argue that the paradox of outside criticism is false because it is based on an ill-conceived notion of scientific community. First, I introduce a distinction between two notions of scientific community, a standard-based and an expertise-based notion. Second, I argue that the paradox of outside criticism presupposes a standard-based notion of scientific community. Third, I argue that a standard-based notion of scientific community is ill-conceived because it does not make room for changes in the standards of argumentation in scientific communities. An expertise-based notion is superior to a standard-based notion because it makes room for changes in the standards of argumentation. Fourth, I argue that an expertise-based notion of scientific community undermines the paradox of outside criticism.

By a standard-based notion of scientific community I mean the view that a scientific community is defined as a group of scientists who share certain standards of argumentation. By an expertise-based notion of scientific community I mean the view that a scientific community is defined as a group of scientists who share an object of inquiry and a particular approach to its study. Scientific communities in the latter sense are mapped on scientific disciplines and specialties. Whereas a standard-based notion can be traced back to Michael Polanyi's 1962 essay "The Republic of Science: Its Political and Economic Theory" (Jarvie 2001) and Thomas Kuhn's 1962 book *The Structure of Scientific Revolutions* (Jacobs 2002), an expert-based notion can be found in the 1969 "Postscript" that Kuhn added to the second edition of *The Structure of Scientific Revolutions* (Kuhn 1996, pp. 176–177). In the "Postscript," Kuhn demands that "scientific communities can and should be isolated without prior recourse to paradigms; the latter can then be discovered by scrutinizing the behavior of a given community's members" (1996, p. 176).

I argue that the paradox of outside criticism presupposes a standard-based notion of scientific community. If we adopt a standard-based notion of scientific community, then two communities are identified as different communities insofar as they have different and nonoverlapping standards of argumentation. Given this notion of scientific community, there seems to be only two alternatives: either two separate communities with no shared standards and no opportunities for outside criticism or two overlapping communities with some shared standards and no opportunities for outside criticism. In the former case, outside criticism is made impossible by lack of shared standards; in the latter case, it is made impossible by lack of outside perspectives.

A contextualist theory of epistemic justification enables us to see why a standard-based notion of scientific community is ill-conceived. Contextualism advises scientists to adopt standards of argumentation as default entitlements and not as plain assumptions. This means that standards of argumentation are open to changes within scientific communities. Such changes may concern how standards are interpreted and applied in specific circumstances or how they are weighed in comparison to other standards. So, if we adopt a contextualist understanding of standards as default entitlements, then scientific communities cannot be identified on the basis of their standards alone. They have to be identified as particular communities on the basis of some other criterion. An alternative criterion is offered by an expertise-based notion of scientific community, which defines communities as groups of scientists who share an object of inquiry and a particular approach to its study.

Also, I argue that an expertise-based notion of scientific community undermines the paradox of outside criticism in two ways. First, if we adopt an expertise-based notion of scientific community, then two communities are identified as different communities in virtue of having different domains of expertise and not in virtue of having different standards of argumentation. Therefore, being two different communities does not imply that these two communities have different and nonoverlapping standards of argumentation. It is easy to perceive how two communities can find a common ground for argumentation. A common ground can consist of shared standards of argumentation as well as shared empirical beliefs. Also, scientific communities can be related in many other ways besides being either separate or nonseparate. For example, scientific communities can be embedded in each other. Endocrinologists and physiologists may form two different communities, but they collectively belong to a broader community of biomedical researchers. And even if they differ about some standards of argumentation, the background theories and standards of argumentation they share as members of a broader community enable them to exchange arguments and counterarguments. Moreover, even if two communities use different concepts, there is no reason to believe that a translation of these concepts from one conceptual framework to another would be impossible in principle (Davidson 1984). Therefore, it is a mistake to assume that if there are two different communities, then effective and transformative criticism is not possible.[7]

An expertise-based notion of scientific community undermines the paradox of outside criticism in yet another way. The paradox of outside criticism is based on the assumption that as soon as two communities share some standards of argumentation, they are no longer two different communities. However, if two communities are different in virtue of having different domains of expertise, then it is possible to conceive that two communities share some standards of argumentation, and they still remain as two different communities (in virtue of their different domains of expertise). Outside criticism can be understood as a critical perspective coming from another community, either a scientific community or a community of nonscientists. Outside criticism is not by definition criticism based on different standards of argumentation. Therefore, it is a mistake to assume that as soon as two communities share some standards of argumentation there are no longer opportunities for outside perspectives.

To summarize, the paradox of outside criticism is false because it is possible to conceive of different communities having a common ground for argumentation and opportunities for outside perspectives. In other words, the notion of outside criticism is not incoherent. And therefore, it is possible to suggest, as Longino does, that communities are evaluated on the basis of whether they are responsive to outside criticism that satisfied certain criteria. The paradox of outside criticism does not pose a challenge to Longino's social epistemology because it is not a paradox at all.

Does Longino's social epistemology need a naturalistic justification?

Besides the paradox of outside criticism, Longino's social epistemology is challenged by another objection. Miriam Solomon and Alan Richardson (2005) argue that Longino's normative account of objectivity lacks naturalistic justification because Longino does not provide empirical evidence in support of the four norms that constitute her social account of objectivity. They suggest that empirical evidence is needed to justify the claim that the four norms of public criticism, uptake of criticism, shared standards of argumentation, and tempered equality of intellectual authority are both attainable and epistemically desirable, that is, they promote either truth or empirical success. As Solomon and Richardson explain, "Longino rests with presenting her standards as intuitively reasonable" (2005, p. 216; see also Solomon 2001, pp. 143–145). In their view, this is unsatisfactory because "'intuition' has a poor track record in philosophy of science" (2005, p. 213).

I argue that a contextualist theory of epistemic justification provides an epistemic justification for the four norms that constitute Longino's normative account of objectivity. Longino's four norms are justified because they advance epistemic responsibility. Recall that in Longino's account, a scientific community should be characterized by (1) publicly recognized forums for criticism, (2) the uptake of criticism, (3) publicly recognized standards of argumentation, and (4) the equality of intellectual authority (2002, pp. 129– 134; see also 1990, pp. 76–79). The first norm advances epistemic responsibility by guaranteeing that those presenting a challenge to a belief have a hearing. The second norm advances epistemic responsibility by requiring that an appropriate challenge receive a response. Indeed, the second norm is parallel to the notion of defense commitment because the second norm implies that at least some scientists in a community have a duty to defend or revise the community's view if it is challenged in an appropriate way. The third norm advances epistemic responsibility by requiring that the standards that determine what counts as an appropriate challenge be openly communicated to the members of the scientific community as well as to outsiders who wish to challenge a community's view. The fourth norm advances epistemic responsibility by requiring that an appropriate challenge be taken seriously independently of who presents it.

The epistemic justification that contextualism provides for the four requirements is not a naturalistic one. Thus, it may be disappointing for Solomon and Richardson who demand that a normative account of objectivity be given a naturalistic justification. I argue that we should not take this demand at face value because, in some cases, a naturalistic justification is not attainable. There is a reason to believe that it is difficult to provide empirical evidence for the hypothesis that the equality of intellectual authority promotes the epistemic success of scientific research. Gerald L. Geison (1981) has attempted to test claims about leadership styles and scientific success in his comparative study of laboratory-based research schools. Geison has compared nine research schools in order to see whether there is a correlation between scientific success and such factors as charismatic leaders and informal leadership styles (Geison 1981, p. 24). Yet, he hesitates to make any conclusions on the basis of his data because it is difficult to determine whether certain leadership styles have been present in otherwise quite different research schools and because the impact of leadership styles cannot be isolated from the impact of other significant factors such as adequate financial support. Thus, Geison's study raises doubts about whether empirical evidence for a hypothesis concerning the impact of leadership styles on scientific success is attainable.[8]

I suggest that when a naturalistic justification for a normative account of scientific knowledge is not attainable, a normative account can settle for a less ambitious goal. It can aim to tell what an epistemically responsible thing to do is. Longino's normative account of objectivity achieves this aim because the four norms advance epistemic responsibility in scientific communities.

Conclusion

Philosophers of science have not addressed the stakeholder question in science in an adequate way. My explanation for this is that the paradox of outside criticism prevents philosophers from seeing the stakeholder question as a philosophical problem in its own right. The paradox of outside criticism is that if the so-called outside criticism is to be effective criticism, then it is not outside criticism at all; it is more appropriately called inside criticism. Given the paradox of outside criticism, the stakeholder question appears to be a nonissue. There is no need to ponder the role of outsiders in epistemic justification because the so-called outsiders are expected to become insiders insofar as epistemic justification is concerned.

I have argued that the paradox of outside criticism is false because it is possible to conceive of different communities having a common ground for argumentation and opportunities for outside perspectives. An implication of this result is that the stakeholder question cannot be subsumed under a general theory of the standards of argumentation in the sciences. Such a move would eclipse a number of philosophical questions that deserve attention in their own right: Under what conditions do outside perspectives have epistemic value and why do they have such value? What are the epistemic responsibilities of those who wish to present a challenge to the views of a particular scientific community? What are the epistemic responsibilities of those who encounter challenges from outsiders, either laypersons or scientists from other disciplines and specialties?

I have discussed Longino's social epistemology as a representative of a stakeholder theory of scientific knowledge. I have argued that a contextualist theory of epistemic justification provides an epistemic justification for the claim that outsiders to particular scientific communities can legitimately intervene in scientific debates. I have also defended Longino's social epistemology against the objection that it lacks naturalistic justification. Whereas Solomon and Richardson (2005) claim that empirical evidence is required to show that the four norms in Longino's normative account of objectivity are attainable and desirable, I have argued that Solomon's and Richardson's naturalistic standards of justification for a normative theory of scientific knowledge may not be attainable. I have suggested that, when a naturalistic justification is not attainable, a normative theory of scientific knowledge can aim to provide another kind of justification. It can aim to show that certain norms advance epistemic responsibility.

This said, it remains to be clarified how a stakeholder theory of scientific knowledge can proceed in order to understand the epistemic value of outside perspectives and the epistemic responsibilities of parties in cross-community controversies. I suggest that a stakeholder theory of scientific knowledge can aim to analyze case studies of cross-community debates, including those that have led to a transformation in either party's views and those that have not led to any transformation in either party's views. Such case studies can be divided further into two categories: cross-community controversies between scientific communities and between a scientific community and a community of laypersons. The advantage in analyzing case studies is that they enable one to give content to the notion of outside perspective. In order to make sense of the notion of outside perspective, we need to specify "perspective on what" (Rolin 2006, p. 132).

An example of such a case study is Brian Wynne's analysis of the conflict between sheep farmers and government scientists in Northern England in the aftermath of the Chernobyl accident in 1986. Wynne argues that sheep farmers' knowledge about sheep behavior and local environmental conditions was relevant for scientists' attempts to understand and explain the consequences of the radioactive fallout (1992, p. 287). Also, Wynne argues that scientists failed to pay attention to sheep farmers' "lay expertise" because scientists lacked the means to reflect on the limits of their scientific expertise (1992, p. 298). In terms of the contextualist theory of epistemic justification I have introduced here, the problem revealed in Wynne's case study is that scientists did not realize that one does not need to be an expert in order to raise an appropriate challenge to scientists' assumptions. Scientific knowledge is, to a large extent, arcane; however, because of the complex web of background assumptions supporting evidential reasoning, scientific knowledge can include assumptions that are accessible to laypersons. In order to raise an appropriate challenge, it is sufficient to give a reason to suspect that an assumption is false. Thus, the standards for making an appropriate challenge are often lower than the standards for making a scientific contribution in a specialty.[9]

Collins and Evans suggest that the communication of relevant knowledge from sheep farmers to the scientific community could have been facilitated by an "interactional expert" (2002, p. 254). An interactional expert is a person who knows enough of a scientific specialty in order to be able to translate an outside perspective into the language of the specialty and to construct arguments that appeal to the standards of argumentation accepted in the target community (2002, pp. 255–256). However, an interactional expert does not need to be a member of the target community. Thus, Collins and Evans suggest that the scientific community as a whole could make epistemically valuable use of experts whose specialty is to mediate between scientific specialties.

In this chapter, I have argued that the paradox of outside criticism has to be dissolved before we can even begin to ask questions about the epistemic role of interactional experts in science. If one accepts the paradox of outside criticism, then it is difficult to understand how an interactional expert could accomplish anything at all. Why would the target community care about criticism coming from another community with different standards of argumentation? However, if we reject the paradox of outside criticism as false, then it is possible to understand how an interactional expert can accomplish her work successfully. An interactional expert can seek a common ground for argumentation, and she can articulate an outside perspective in such a way that it appeals to the standards of the target community. This work may not always be easy (intellectually, socially, and financially) but there is no reason to believe that it is impossible.

Acknowledgments

I wish to thank Jeroen Van Bouwel, Harry Collins, Endla Lõhkivi, Martin Kusch, and K. Brad Wray for their comments on earlier versions of this manuscript. Financial support for this work has been provided by the Academy of Finland.

Notes

1. The term *stakeholder* has been introduced in business ethics as a deliberate play on the term *shareholder* to suggest that there are other parties having a "stake" in the decision-making of the modern corporation in addition to those holding shares (Freeman 2002).
2. The term *layperson* should be used with caution in this context because scientists are laypersons in relation to some other scientists' disciplines and specialties.
3. John Hardwig (1991) and Philip Kitcher (1993) focus on relations of trust among scientists, Miriam Solomon (2001) on distribution of research effort within scientific communities, David Hull (1988) on competition among scientists, and K. Brad Wray (2002 and 2006) on collaboration among scientists. An exception to the tendency to focus on social relations within scientific communities is found in the works of those philosophers who argue that scientists need to pay attention to outside perspectives in order to be able to define an acceptable level of risk in the assessment of evidential warrant (see e.g., Douglas 2004; Lacey 1999; Mayo and Hollander 1991).
4. Longino's social epistemology is akin to Steve Fuller's (2000) view in that it emphasizes the scientists' responsibilities towards stakeholders. However, Longino's and Fuller's views differ in an important respect. Fuller introduces the controversial idea that science policy makers should enforce scientists to justify their claims to stakeholders on a regular basis by organizing open forums for such debates (2000, p. 142). Longino suggests merely that scientists have a duty to respond to stakeholder criticism insofar as such criticism is raised and it satisfies certain criteria.
5. Elizabeth Anderson (2006) argues that a diversity of perspectives is of epistemic value especially when social scientists study complex social phenomena. This is because information that is relevant for understanding complex social phenomena is typically dispersed across society and distributed asymmetrically depending on individuals' geographic location, social class, occupation, education, gender, race, age, and so forth (see also Bohman, 2006).
6. Philip Kitcher (1993) and Miriam Solomon (2001) argue that cognitive diversity is of epistemic value because it generates and maintains an efficient distribution of research effort. Yet their ideas about how cognitive diversity is best brought about are different from Longino's. Kitcher assumes that there are sufficient incentives for scientists to pursue nonconventional and risky lines of research because scientific communities value originality (1993, p. 352). Thus, he does not make any suggestions as to how to improve the institutional arrangements of science. Solomon wishes to make room for cognitive diversity by introducing the controversial claim that nonepistemic factors should be allowed to enter into individual scientists' decisions about what theories to work with (2001, p. 135). Solomon recommends also that science policy makers pay attention to the distribution of

research effort (2001, p. 150). Neither Kitcher nor Solomon addresses the question of who are recruited into the profession. And neither one of them addresses the question about the roles stakeholders are allowed to play in epistemic justification. For Longino, these two questions are crucial in understanding how cognitive diversity is best generated and maintained in scientific communities.

7. I wish to thank K. Brad Wray for drawing my attention to complex relations between scientific communities.

8. I wish to thank Martin Kusch for drawing my attention to Gerald Geison's study (1981) and other historical studies on research schools (see also Kusch 1999, pp. 248–249).

9. A nonscientist's lay expertise is sometimes called "experience-based expertise" (Collins and Evans 2002) or "citizenry personal expertise" (Morgan 2006). Mary Morgan (2006) argues that laypersons are experts in the specific circumstances of their life situations simply because they have access to firsthand and very close secondhand experience of these circumstances.

References

Anderson, Elizabeth (2006) 'The Epistemology of Democracy.' *Episteme* 3(1–2), 8–22.

Bohman, James (2006) 'Deliberative Democracy and the Epistemic Benefits of Diversity.' *Episteme* 3(3), 175–91.

Collins, Harry and Robert Evans (2002) 'The Third Wave of Science Studies: Studies of Expertise and Experience.' *Social Studies of Science* 32(2), 235–96.

Crasnow, Sharon (2003) 'Can Science Be Objective? Feminism, Relativism, and Objectivity.' In: *Scrutinizing Feminist Epistemology: An Examination of Gender in Science*, ed. Cassandra L. Pinnick, Noretta Koertge and Robert F. Almeder. New Brunswick, NJ: Rutgers University Press, pp. 130–41.

Davidson, Donald (1984) 'On the Very Idea of a Conceptual Scheme.' In: *Inquiries into Truth and Interpretation*. Oxford: Oxford University Press, pp. 183–98.

Douglas, Heather (2004) 'Border Skirmishes between Science and Policy.' In: *Science, Values, and Objectivity*, ed. Peter Machamer and Gereon Wolters. Pittsburgh, PA: University of Pittsburgh Press, pp. 220–44.

Epstein, Steven (1996) *Impure Science: AIDS, Activism and the Politics of Knowledge.* Berkeley: University of California Press.

Freeman, R. Edward (2002) 'Stakeholder Theory of the Modern Corporation.' In: *Ethical Issues in Business: A Philosophical Approach*, 7th ed, ed. Thomas Donaldson, Patricia H. Werhane and Margaret Cording. Upper Saddle River, NJ: Prentice Hall, pp. 38–48.

Fuller, Steve (2000) *The Governance of Science: Ideology and the Future of the Open Society.* Buckingham, UK: Open University Press.

Geison, Gerald L. (1981) 'Scientific Change, Emerging Specialties, and Research Schools.' *History of Science* 19, 20–40.

Goldman, Alvin I. (2002) 'Experts: Which Ones Should You Trust?' In: *Pathways to Knowledge: Private and Public*. Oxford: Oxford University Press, pp. 139–63.

Hardwig, John (1991) 'The Role of Trust in Knowledge.' *Journal of Philosophy* 88, 693–708.

Hempel, Carl (1965) 'Science and Human Values.' In: *Aspects of Scientific Explanation and Other Essays in the Philosophy of Science*. New York: Free Press, pp. 81–96.

Hull, David (1988) *Science as a Process: An Evolutionary Account of the Social and Conceptual Development of Science*. Chicago: University of Chicago Press.

Jacobs, Struan (2002) "The Genesis of 'Scientific Community'." *Social Epistemology* 16(2), 157–68.

Jarvie, Ian (2001) 'Science in a Democratic Republic.' *Philosophy of Science* 68(4), 545–64.

Jasanoff, Sheila (2004) 'Science and Citizenship: A New Synergy.' *Science and Public Policy* 31(2), 90–94.

Jasanoff, Sheila (2005) 'Science and Environmental Citizenship.' In: *Handbook of Global Environmental Politics*, ed. Peter Dauvergne. Northampton, UK: Edward Elgar, pp. 365–82.

Kitcher, Philip (1993) *The Advancement of Science: Science without Legend, Objectivity without Illusions*. Oxford: Oxford University Press.

Kitcher, Philip (2001) *Science, Truth and Democracy*. New York: Oxford University Press.

Kuhn, Thomas (1996) *The Structure of Scientific Revolutions*, 3rd ed. Chicago: University of Chicago Press.

Kusch, Martin (1999) *Psychological Knowledge: A Social History and Philosophy*. London: Routledge.

Lacey, Hugh (1999) *Is Science Value Free? Values and Scientific Understanding*. London: Routledge.

Leach, Melissa, Ian Scoones and Brian Wynne, eds (2005) *Science and Citizens: Globalization and the Challenge of Engagement*. London: Zed Books.

Longino, Helen (1990) *Science as Social Knowledge*. Princeton, NJ: Princeton University Press.

Longino, Helen (2002) *The Fate of Knowledge*. Princeton, NJ: Princeton University Press.

Mayo, Deborah and Rachelle D. Hollander, eds (1991) *Acceptable Evidence: Science and Values in Risk Management*. Oxford: Oxford University Press.

Mill, John Stuart (1962) 'On Liberty.' In: *Utilitarianism, On Liberty, and Essays on Bentham*, ed. Mary Warnock. New York: A Meridian Book, pp. 126–250.

Mirowski, Philip (2004) 'The Scientific Dimensions of Social Knowledge and Their Distant Echoes in 20th Century American Philosophy of Science.' *Studies in History and Philosophy of Science* 35, 283–326.

Morgan, Mary (2006) 'Facts of Expertise and Facts of Experience.' Paper presented at the *Social Sciences and Democracy: A Philosophical Perspective* conference in Gent, Belgium, 2006.

Rolin, Kristina (2006) 'The Bias Paradox in Feminist Standpoint Epistemology.' *Episteme* 3(1–2), 125–36.

Rolin, Kristina (forthcoming) 'Contextualism in Feminist Epistemology and Philosophy of Science.' In: *Feminist Epistemology and Philosophy of Science: Power in Knowledge*, ed. Heidi Grasswick. Dordrecht, The Netherlands: Kluwer Academic Publishers.

Solomon, Miriam (2001) *Social Empiricism*. Cambridge, MA: MIT Press.

Solomon, Miriam and Alan Richardson (2005) 'A Critical Context for Longino's Critical Contextual Empiricism.' *Studies in History and Philosophy of Science* 36, 211–22.

Thagard, Paul (1999) *How Scientists Explain Disease*. Princeton, NJ: Princeton University Press.

Turner, Stephen (2006) 'What Is the Problem with Experts?' In: *The Philosophy of Expertise,* ed. Evan Selinger and Robert P. Crease. New York: Columbia University Press, pp. 159–86.

Williams, Michael (2001) *Problems of Knowledge: A Critical Introduction to Epistemology.* Oxford: Oxford University Press.

Wray, K. Brad (2002) 'The Epistemic Significance of Collaborative Research.' *Philosophy of Science* 69(1), 150–68.

Wray, K. Brad (2006) 'Scientific Authorship in the Age of Collaborative Research.' *Studies in History and Philosophy of Science* 37, 505–14.

Wynne, Brian (1992) 'Misunderstood Misunderstanding: Social Identities and Public Uptake of Science.' *Public Understanding of Science* 1, 281–304.

Part II

The Social Sciences Improving Democratic Theory and Practice

4
Improving Democratic Practice: Practical Social Science and Normative Ideals

James Bohman

Democracy can be examined in a variety of ways in the social sciences. First, it can be examined objectively, so that the social scientist seeks to explain the operation and impact of various features of democracy, such as the different ways in which political life can be organized, the effects of political parties on legislation, the voting behavior of various groups, and so on. One outcome of such inquiry may be various generalizations about democracy, concerning parliamentary systems, voting schemes, or the tendencies of a democracy to go to war or to prevent famines. Social scientists can also take a practical orientation and seek not just to explain or interpret what democracy is, but to change it. In his essay "Ideal Understanding," Martin Hollis, a social scientific defender of the Enlightenment, links this sort of social science to the analysis of the agency of knowledgeable social actors. "Actors," he remarks, "have natural, social and rational powers" (Hollis 1977, p. 180). Hollis goes on linking reason and freedom to specifically social and normative powers and capabilities that make it possible for an actor to become an agent who shapes the social world. This idea of freedom and powers might also be the basis of a kind of social science that aims at understanding the conditions for the proper exercise of freedom and agency. Understood in this way, the social sciences are indeed "moral sciences" in the Millian sense, and pragmatism and Critical Theory offer some of the most developed philosophical justifications and analyses of such a practical approach. In the context of their accounts of agency, the ideal of a robust democracy as a means to achieve freedom has long been the focus of such practical social science.

At the institutional level, democracy answers some of the moral and political problems raised by such uses of social science. One suspicion of the Enlightenment, articulated by its critics, is that it gives to its disciples the power to bring about a rational order. But this is to see the problem of enlightenment as an *engineering problem*, in which the social scientifically informed experts alone possess the knowledge necessary to make optimal choices. In his critique of Enlightenment cosmopolitanism, for example,

Stephen Toulmin argues that its conception of rationality is committed to a "central apex of power" (Toulmin 1990, p. 209).[1] If we think of such a project instead as a democratic one, then such power must be dispersed, and experts are only one sort of participant in deliberative inquiry into solutions to problems and the correct rules and laws that promote justice. Similarly, democratic rules are both "enabling and constraining," so that they are judged not simply by what their constraints achieve instrumentally in protecting individual rights, but also in terms of the active processes of shaping and interpreting these rules and laws in an ongoing way. Finally, democracy is an institution that is not merely universal: it is also realized in particular polities in social space and historical time that are always parochial to some extent. Thus, even if democracy solves the practical problem for purely instrumental social science by providing the means and location for the collective realization of various moral and political aims, it is also a practice that is itself in need of ongoing improvement so that it better achieves its intrinsic and instrumental aims and ends, including intrinsic aims such as self-rule and instrumental aims such as providing citizens with the means to avoid the great harms of political power, famine, and war.

Such a practical emphasis of the moral and political significance of realizing democratic ideals produces an equal and opposite reaction: skepticism about whether democracy is indeed such a means and location for realizing the goods at which social scientific reformers aim. Some of these skeptical accounts are themselves practical, even if in a negative sense: they aim at uncovering the obstacles to democracy under current circumstances, such as complexity, mass society, human irrationality, or globalization. They count as practical forms of inquiry because they help to develop or develop themselves alternatives to democratic practice that improve it by changing it. That such accounts are skeptical in a practical sense can be seen in the ways in which they lead to advocating that democracy be replaced by alternative forms of rule, as Lippmann did in proposing technocracy as a solution to the invincible irrationality of mass rule. By identifying obstacles that democracy must overcome, such critics invite practical reinterpretations of democracy that address such problems, as Dewey did in his analysis of the emergence of new publics.

My goal here is to show the multiple ways in which social science can serve the practical end of improving democracy. The social scientific tools for improving democracy are varied: generalizations about features and outcomes of democratic practice, social facts in the pragmatic sense of obstacles and resources, responses to skepticism (or negative facts) about human reasoning, and the analysis of failures of democratic practice. However, no specific methods and theories can lay claim to having a special status in this regard. Rather, it is in the role of critic or reformer that they are put to practical use in the wider context of the purposes of a progressive or ameliorative politics. By whatever methods they employ, the social sciences are practical only if they develop a form of inquiry that shows how it is feasible to realize certain political ideals under contemporary circumstances of politics. I will focus on forms of social science inspired by pragmatism and Critical Theory, both of which emphasize *praxis*, the form of social activity aimed at the realization of norms and ideals in interaction and intersubjective processes—"praxeology" in Andrew Linklater's sense (Linklater 2001, pp. 38–41). The ideal in question for such approaches is a robust and deliberative form of democracy, the institutions of which also provide the basis for the realization of ideals of freedom and equality. At the same time, social facts can also outstrip particular institutional realizations of these ideals. Social science thematizes just these problematic features as both obstacles and opportunities for the further development of the democratic ideal, including globalization, war and its tendencies for domination at home and abroad, and the need to make decisions and promote good reasoning under conditions of complexity without the appropriate institutional structure. If it is merely skeptical, social science cannot solve such problems; if it ignores social facts, it cannot meet the need to improve democratic practice.

Social facts and normative ideals

In what respect should a practical and critical social science aimed at improving or even transforming democracy be concerned with social facts? A social scientific praxeology understands facts in relation to human agency rather than independent of it. Pragmatic social science is concerned not merely with elaborating an ideal in convincing normative arguments, but also with its realizability and its feasibility. In this regard, any political ideal must take into account general social facts if it is to be feasible; but it must also be able to respond to a series of social facts and thus to skeptical challenges that suggest that current circumstances make such an ideal impossible. With respect to democracy, these facts include expertise and the division of labor, cultural pluralism and conflict, social complexity and differentiation, globalization and the fact of increasing social interdependence, to name a few. In cases where "facts" challenge the very institutional basis of modern political integration, normative practical inquiry must seek to extend the scope of political possibilities rather than simply accept the facts as fixing the limits of political possibilities once and for all. For this reason, social science is practical to the extent that it is able to show how political ideals that have informed these institutions in question are not only still possible, but also feasible under current conditions or modification of those conditions. As I have been arguing, the ideal in question for pragmatism and recent critical social theory inspired by pragmatism is a robust and deliberative democracy—precisely as a key aspect of a wider historical ideal of human emancipation and freedom from domination.

Not all constraints can be thought of in terms of such large-scale and long-term social facts. Democracy also requires voluntary constraints on action, such as commitments to basic rights and constitutional limits on political power that make it possible. Social facts, on the other hand, are non-voluntary constraints that affect the scope of application of democratic principles. Taken up in a practical social theory oriented to changing the ideal of democracy to make it more robust, social facts no longer operate simply as constraints. For Rawls, "the fact of pluralism" (or the diversity of moral doctrines in modern societies) is just one such permanent feature of modern society that is directly relevant to political order, because its conditions "profoundly affect the requirements of a workable conception of justice" (Rawls 1999, p. 424). However, as facts such as pluralism are made permanent in those modern institutions and ideals developed after the Wars of Religion, Rawls sees constitutional democracy and freedom of expression as promoting rather than inhibiting their development. This fact of pluralism thus alters how we are to think of the *feasibility* of a political ideal, but it does not touch on its realizability or possibility. Similarly, Rawls's "fact of coercion," understood as the fact that any political order created around a single doctrine would require oppressive state power, concerns not the realizability or possibility of a particular ideal, but its feasibility as a "a stable and unified order," under the conditions of pluralism (Rawls 1999, p. 425). Thus, for Rawls, regardless of whether they are considered in terms of possibility or feasibility, they are only considered as *constraints*—as restricting what is politically possible or what can be brought about by political authority and power. In keeping with the nature of pluralism, not all actors and groups experience the constraints of pluralism in the same way. It may be experienced by some actors as a hurdle in their attempt to realize a democracy based on a social consensus of shared values and norms; for others, it may suggest that we ought not to think of a common identity as a requirement for democracy at all. Different perspectives on the fact of pluralism result not only from occupying different places in the social relations among groups, but also from the determination of the political ideal that such actors may seek to realize.

If this were the only role of facts in Rawls's political theory, then it would not be a full practical theory in the sense that I am using the term here. Rawls's contribution is that social facts differ in kind, maintaining that social facts such as the fact of pluralism are permanent and not merely to be considered in narrow terms of functional stability. Social facts related to stability may indeed constrain feasibility without being limits on the possibility or realizability of an ideal as such; in the case of pluralism, for example, democratic political ideals other than liberalism might be possible. Without locating a necessary connection between its relations to feasibility and possibility, describing a social fact as "permanent" is not entirely accurate. It is better instead to think of such facts as "institutional facts" that are deeply entrenched in some historically contingent, specific social order rather than as necessary outcomes of free societies as such. Thus, what Rawls calls "permanent" facts about modern societies are rather those determinations that are embedded in relatively long-term social processes, the consequences of which cannot be reversed in a short period of time—such as a generation—by political action. Practical theories thus have to consider the ways in which such facts become part of such a process, which might be called "generative entrenchment" (Wimsatt 1974). By "generative entrenchment of social facts," I mean that the relevant democratic institutions promote the very conditions that make the institutional social fact possible in assuming those conditions for their own possibility. When the processes at work in the social fact then begin to outstrip particular institutional feedback mechanisms that maintain it within the institution, then the institution must be transformed if it is to stand in the appropriate relation to the facts that make it feasible and realizable. All institutions, including democratic ones, entrench some social facts in realizing their possibility.

Consider Habermas's similar use of social facts with respect to this. As with Rawls, for Habermas, pluralism and the need for coercive political power make the constitutional state necessary, so that the democratic process of law making is governed by a system of personal, social, and civil rights. However, Habermas introduces a more fundamental social fact for the possibility and feasibility of democracy: the structural fact of social complexity. Complex societies are "polycentric," with a variety of forms of order, some of which, such as nonintentional market coordination, do not answer to the ideals of democracy. This fact of complexity limits political participation and changes the nature of our understanding of democratic institutions. Indeed, this fact makes it such that the principles of democratic self-rule and the criteria of public agreement cannot be asserted simply as the proper norms for all social and political institutions. As Habermas puts it, "Unavoidable social complexity makes it impossible to apply the criteria [of democratic legitimacy] in an undifferentiated way" (1996, p. 305). This fact makes a certain kind of structure ineluctable because complexity means that democracy can "no longer control the conditions under which it is realized." In this case, the social fact has become "unavoidable," and certain institutions are necessary for the social integration for which there is "no feasible alternative" (Habermas 2001, p. 122). Taken as a macro-sociological fact about modern societies, complexity may indeed make impossible any direct realization of democracy as a single organizing principle for all social institutions. However, appealing to this fact says little about the field of indirect and institutionally mediated institutional designs that are still feasible. These mediated forms of democracy would in turn affect the conditions that produce social complexity itself. Complexity may appear to be merely a technical problem, mastered by institutions equipped with legitimate authority and expert knowledge. However, this solution assumes that the fact of social complexity is then not the same across all feasible institutional realizations of democracy, and some ideals of democracy may rightly encourage the preservation of aspects of complexity, such as the ways in which the epistemic division of labor may promote wider and more collaborative problem solving and deliberation on ends.

When seen in light of the requirements of practical social science, constructivists are right to emphasize how agents continually act so as to produce and maintain social factual realities, even if they do not do so necessarily under conditions of their own making. In this context, an important contribution of pragmatism is precisely its interpretation of the practical status of social facts. Thus, Dewey sees social facts always related to "problematic situations," even if these are more felt or suffered than fully recognized as such. The way to avoid turning problematic situations into empirical-normative dilemmas is, as Dewey suggests, seeing facts themselves as practical: "[f]acts are such in a logical sense only as they serve to delimit a problem in a way that affords indication and test of proposed solutions" (1986, p. 499). They may serve this practical role only if they are seen in interaction with our understanding of the ideals that guide the practices in which such problems emerge, thus, where neither fact nor ideal is fixed and neither is given justificatory or theoretical priority. In response to Lippmann's insistence on the pre-eminence of expertise, Dewey criticized the possibilities inherent in "existing political practice, [which] with its complete ignoring of occupational groups and the organized knowledge and purposes that are involved in the existence of such groups, manifests a dependence upon a summation of individuals quantitatively" (Dewey 1991, pp. 50–51). In reply to Lippmann's elitist view of majority rule, Dewey held on to the possibility and feasibility of democratic participation by the well-informed citizen. However, he recognized that existing institutions were obstacles to the emergence of such a form of participatory democracy in an era when "the machine age has enormously expanded, multiplied, intensified and complicated the scope of indirect consequences" of collective action and where the collectives—affected by actions of such a scope—are so large and diverse "that the resultant public cannot identify and distinguish itself" (Dewey 1988, p. 255 and p. 314). Our democratic ideals have been shaped by outdated "local town meeting practices and ideal[s]," even as we [Americans] live in a "continental nation state" whose political structures encourage the formation of "a scattered, mobile and manifold public" (Dewey 1988, pp. 280–81) that has yet to recognize itself as a public and to form its own distinct common interests. Thus, Dewey saw the solution in a transformation both of what it is to be a public and of the institutions with which the public interacts. Such interaction will provide the basis for determining how the functions of the new form of political organization will be limited and expanded, the scope of which is "something to be critically and experimentally determined" in democracy as a mode of practical inquiry about social facts (Dewey 1988, p. 281).

The pragmatist conception thus clearly suggests an understanding of social facts for a social scientific praxeology, namely, that it is better to think of social facts as creating a new space of democratic possibilities, opening up some while foreclosing others. However, while accepting that, Critical Theory insists that such a praxeology may lead to judgments about the need for structural transformation of democracy as an ideal and as a set of institutions. "Given the limits and possibilities of our world," Dahl asks, "is a third transformation of democracy a realistic possibility?" (1989, p. 224). This challenging situation could very well call into question the current normative conception of democracy itself; perhaps its array of possibilities is too restricted or forces too many hard choices upon us. The question is not just one of current political feasibility, but also of possibility, given that we want to remain committed in some broad sense to democratic principles of self-rule even if not to the set of possibilities provided by current institutions. How do we identify such fundamentally unsettling facts? I turn next to the discussion of a specific social fact, the "fact of globalization," and interpret it not as a uniform process but as a problematic situation that both is experienced in different and even contradictory ways from a variety of perspectives and is differently assessed with respect to different normative ideals of democracy.

The fact of globalization and multiperspectival democracy

For some, the fact of globalization permits a direct inference to the need for new and more cosmopolitan forms of democracy and citizenship. Whatever the specific form they assume, the usual arguments for political cosmopolitanism are relatively simple despite the fact that the social scientific analyses employed in them are highly complex and empirically differentiated in their factual claims. In discussions of theories of globalization, the fact of global interdependence refers to the unprecedented extent, intensity, and speed of social interactions across borders, encompassing diverse dimensions of human conduct from trade and cultural exchange to migration.[2] The inference from these facts of interdependence is that existing forms of democracy within the nation-state must be transformed and that institutions ought to be established that solve problems that transcend national boundaries.[3] Here globalization is taken to be a macro-sociological, aggregative fact that constrains the realization of democracy without further political integration and congruence, much in the same way that Habermas sees the fact of social complexity as constraining the possible realizations of democratic principles and thus making representation, voting, and bureaucracy necessary features of modern societies, democratic or not. The Deweyan alternative is to see that facts "have to be determined in their dual function as obstacles and as resources," as multidimensional in such a way as to hold out the possibility of transforming the situation by transforming democracy itself (Dewey 1986, pp. 499–500). The "mere" fact of the wider scale of interaction is thus inadequate on its own and does not capture what role globalization may play as a problematic situation for the emergence of new democratic possibilities.

A pragmatic interpretation of social facts then encourages us to see globalization as both a resource and an obstacle for democracy. While many social scientific accounts of globalization suggest that it is a single, unified process, a pragmatic analysis sees it as multidimensional, constraining democracy in some respects and promoting it in others. The interdependencies it produces sometimes have positive consequences to be promoted and sometimes negative ones to be avoided. What is important here, as in the case of the fact of pluralism, to see that this process can be experienced in different ways by different peoples or political communities, given that it is a multifaceted and multidimensional process producing "differential interconnectedness in different domains" (Held et al. 1999, p. 27). In some domains such as global financial markets, globalization is profoundly uneven and deeply stratified, reinforcing hierarchies and disproportion. Inequalities of access to and control over aspects of globalizing processes may reflect older patterns of subordination and order, even while the process produces new ones by excluding some communities from financial markets and by making others more vulnerable to its increased volatility (Ibid, p. 213).[4] The fact of globalization is a new sort of social fact whose structure of enablement and constraint is not easily captured at the aggregative level. It is even experienced in contradictory ways looking at its consequences and impacts, which differ across various domains and at various locations.

How can we understand its practical significance in a normative theory concerned with the feasibility and realization of the democratic ideal? Given the sort of fact that globalization is, there are a number of possible responses to it. The lack of effective global institutions suggests that many of its practical consequences can be seen as suboptimal outcomes of collective coordination problems, solved to the mutual benefit of all by the appropriate application of technical knowledge or expertise. This understanding of the fact of globalization leads, for example, to investing greater authority in international financial institutions in which experts are guided by economic theories of proper market functioning that suggest the policies necessary to avoid instability and volatility in global financial markets. While admitting that such theoretical knowledge has its role to play, Critical Theory suggests that such an understanding of the problematic situation of globalization as a coordination problem ignores the way in which such institutions create potential for domination. The exclusion of nonexperts leads to policies that exclude valuable perspectives. As Bina Agarwal points out, the systematic exclusion of the perspective of women from deliberation on Community Forestry groups in India and Nepal leads to inferior practice (see Agarwal 2001). Because women had primary responsibility for wood gathering in their search for cooking fuel, they possessed greater knowledge of what sort of gathering was sustainable and about where trees were that needed protection. Mixed groups of guards thus would provide a much more effective method of enforcement and an epistemically superior implementation of these ends. This requires a kind of second-order testing and accountability not yet available in international financial institutions because they do not yet fully consider, among other things, the social disintegration that globalization may bring or the local circumstances for policy implementation (see Rodrik 1994; also Woods 2001).

The example of global financial markets raises the question of the sort of practical knowledge employed in institutions and how the method of inquiry in them may promote or inhibit democratic alternatives. The practical alternative to such a solution through *techne* must be multiperspectival, in that it considers all of the variety of experiences and perspectives in a variety of dimensions and domains (Bohman 2002). If globalization is profoundly uneven, then the most interconnected locations and domains of the global structure may be enabled in a variety of ways to achieve various political and economic ends, while those who are less connected may experience the global structure as deeply constraining. In this way, the increased scale and extent of interconnectedness across borders increase the possibilities of domination and, for this reason, require some process of democratization in which the freedom of all may be possible given the consequences of interconnectedness.

For Kant, law is the solution to the problematic situation arising from the new potentials for mutual influence in modernity. Laws are necessary "because individuals and peoples mutually influence each other," either directly or indirectly, under a very specific normative description: "only with regard to the form of the relationship between the two wills, insofar as they are regarded as free, and whether the action of one of the parties can be reconciled with the freedom of the other" (Kant 1970, p. 133, 135). Thus, social interconnectedness points to the possibility of a political structure in which such interaction does not decrease the freedom of all—freedom here understood as freedom from relations of domination that interconnectedness may make possible. Even if we do not accept Kant's inference to the need for global public law, an inference advocated by proponents of cosmopolitan democracy is that normative dimensions of a practical theory of democracy reinforce their importance under the conditions of globalization. That non-domination is a key normative feature of democracy is perhaps the most important contribution of recent Critical Theory, even if its specific analysis emerged in the context of different power relations (Habermas 1974; Habermas 1996; see also Bohman 2002).

One further question about the fact of globalization must be raised in order to understand the inherent possibilities for democracy in it. Is globalization a "permanent" fact for democracy as Rawls described the fact of pluralism for liberalism in that it is deeply embedded in its possible realizations? As many social theorists have argued, globalization is part of long-term social processes beginning in early modernity; as Anthony Giddens put it, "modernity is inherently globalizing" (Giddens 1990, p. 63). Even if it is modernity rather than democracy that is inherently globalizing, then reversing such processes is possible, although not feasible in any short time span and under the normative constraints of democracies and their social preconditions. That is, so long as "globalizing" societies are democratic, we can expect such processes and their impacts to continue. This is not to say that globalization in its current form is somehow permanent or unalterable if we want to realize democratic ideals. Indeed, just how globalization will continue and under what legitimate normative constraints become the proper questions for democratic politics, as citizens and public vigorously interact with those institutions that make globalization a deeply entrenched and temporally stable social fact. Reversing globalization rather than restructuring the institutions that maintain it will likely dramatically undo important normative features of democracy as well. In the political activity attending issues related to globalization, the currently existing yet overly weak normative constraints will be reconstituted along with the institutions themselves. The social fact of globalization is open to democratic reconstruction, should creative reinterpretation of democracy come about. In the face of this fact, the normative emphasis on democracy as a legitimate form of political non-domination suggested by the ideal of human emancipation brings out new features of its ideal. We could call this ideal that of a "multiperspectival polity." Contrary to some interpretations of cosmopolitan democracy, such a polity is not oriented to an ideal according to which democracy means merely that "citizens should be able to freely choose the conditions of their own association,"[5] but rather with establishing political relationships without domination or subordination that could be accepted by everyone. In the next section, I turn to social scientific generalizations about the absence of two indicators of political domination—war and famine—and to explanations of their relative absence in democracies.

No wars and famines: The scope of two generalizations about democracy

There are two main social scientific generalizations about the beneficial effects of democracy, both of which concern what might be thought of as negative facts: the first is that there has (almost) never been a famine in a democracy; and the second is that democracies have (almost) never gone to war with each other. These facts show that the relative absence of two great causes of human suffering—war and famine—can be tied to the operation of distinctive features of democracy. Without some fine-grained explanation of the mechanisms behind them, there is no reason to believe that these generalizations have always held or will always hold in the future, especially if the causes of famine and war are always changing and sometimes are brought about by democratic institutions themselves. Both generalizations have been hotly disputed, leading their defenders to introduce more and more *ceteris paribus* clauses to limit their scope. For example, Bruce Russett has argued that the generalizations have only held since the first half of the twentieth century, given the relative paucity of democratic states before then.[6] Yet, even with such *ceteris paribus* clauses, different mechanisms may do the explanatory work in the cases of famine and war.

Sen's analysis of the relation between famines and democracy begins with two striking facts. The first is that they "can occur even without any decline in food production or availability" (Sen 1999, chapter 5). Even when this is the case, Sen argues that more equitably sharing the available domestic supply is nearly always an effective remedy to get beyond the crisis. Indeed, famines usually affect only a minority of the population of any political entity, and Sen's hypothesis is that their vulnerability to starvation is explained by the loss of certain powers and entitlements that they had before the crisis. The second striking fact goes some way in this direction by showing that when food shortages do occur, they do not have the same disastrous consequences. These facts yield the robust generalization that "there has never been a famine in a functioning multiparty democracy," so that we may conclude that "famines are but one example of the protective reach of democracy" (Sen 1999, p. 184). It would be tempting to associate this sort of security with the achievement of various instrumental freedoms or with one's status as a subject or client of a state or similar institution with an effective and well-funded administration. But even in the case of the protective function of the state, much more is required of democracy to create (or sustain in a crisis) the conditions of entitlement and accountability, as well as the reflexive capacity to change the normative framework. Once the explanation is put in the normative domain, so is the practical understanding of remedies and solutions.

The practical effects of democracy are not directly tied to more effective administrative institutions or even to the consistent application of the rule of law, both of which democracy may achieve. As Sen notes, there are limits to legality: "other relevant factors, for example market forces, can be seen as operating *through* a system of legal relations (ownership rights, contractual obligations, legal exchanges, etc.). In many cases, the law stands between food availability and food entitlement. Starvation deaths can reflect legality with a vengeance" (1986, pp. 165–66). In this sense, the presence of famine must also be explained via the operation of social norms conjoined with the lack of effective social freedom of citizens with regard to their content. The deplorable treatment of native populations in famines caused by colonial administrators has often been due to domination, manifested in their lack of substantive freedoms such as free expression or political participation.

Thus, famine prevention can be gained through fairly simple democratic mechanisms of accountability such as competitive elections and a free press that distribute effective agency more widely than in their absence.

Sen clearly goes further and sees democracy as more than a protective mechanism that can empower certain agents to act and thus enable them to defend the entitlements of citizens. It is also more active and dynamic, offering genuine opportunities to exercise substantial freedoms, including the capability not to live in severe deprivation or to avoid the consequences of gender norms for overall freedom. It is clear that such substantive freedoms depend on normative powers and the emergence of practices of deliberation in which citizens exercise them. For example, India's success in eradicating famines is not matched in areas that require solving persistent problems such as gender inequality, in which the normative powers necessary for effective agency are differentially distributed. There is certainly no robust empirical correlation between democracy and the absence of these problems; they exist in affluent market-oriented democracies such as the United States. The solution for these ills of democracy is not to discover new and more effective protective mechanisms or robust entitlements, because it is hard for some democracies to produce them. Rather, the solution is, as Sen puts it, "better democratic practice" in which citizens are participants in a common deliberative practice and sufficiently protected and empowered to change the distribution of normative powers and take advantage of improved practices. It is certainly the case for women in many developing states and responsible for the unjust distribution of food within families.

To put it somewhat differently, the issue is not merely to construct a more protective democracy, but to create conditions under which an active citizenry is capable of initiating *democratization*, that is, using their power to extend the scope of democratic entitlements and to establish new possibilities of creative and empowered participation. Democracy is on this view the project in which citizens (and not just the agents for whom they are principals) exercise those normative and communicative powers that would make for better and more just democratic practice. This kind of enabling condition is essential to the explanation of the role of phenomena produced by democracy that serve as Sen's explanation: citizens' powers and entitlements.

The "democratic peace hypothesis" is similar to Sen's generalization about famines in that fairly minimal democratic conditions figure in the explanation of the absence of certain types of wars. The generalization is, however, more restricted in the case of war than famine. Democracies do go to war against nondemocracies, although "almost never" against other democracies. Many explanations have been offered for why this is the case, and many of these do not depend on any transformative effects of democratic institutions other than that they provide channels for influence and the expression of citizens' rational interests and presume amity among democracies across borders as the basis for trust. Seen in light of the explanation of the absence of famines, democracy might reasonably be given a similar, more dynamic, and transformative role than is usually offered: by being embedded in democratic institutions, agents acquire the normative role of citizens and the freedoms and powers that provide means by which to avoid the ills of war.

If this is the explanation of peace, it is important to make clear why war and the preparation for war often have the opposite effects. The institutional capability to wage war increases with the executive and administrative powers of the state, which often bypass democratic mechanisms of deliberation and accountability and thus work against democratization (where this is understood precisely as the widening and deepening of the institutional powers of citizens to initiate deliberation and participate effectively in it). At the same time, participating in national self-defense has often been accompanied by the emergence of new rights or their broader attribution to more of the population. Charles Tilly has argued that warfare may have historically been an important mechanism for the introduction of social rights, as the state became more and more dependent on the willingness of citizens to accept the obligations of military service (Tilly 1990). As modern warfare became increasingly lethal and professionalized, however, the institutional powers of the state have outstripped this and other democratic mechanisms. The institutionally embedded normative powers of citizens are no longer sufficient to check the institutional powers of states to initiate wars, and these arrangements have left citizens vulnerable to expanding militarization that has weakened these same entitlements. A new dialectic between the capacities of citizens and the instrumental powers of states has not yet reached any equilibrium, so that there has now emerged a strong negative influence on democratic practices and human rights generally because of the use of state force for the sake of security. Liberal democracies have not only restricted some civil rights, but have become human rights violators, with the use of extralegal detention centers and torture in order to achieve security. As such, they might be said to have become less democratic, certainly in the active sense of creating enabling conditions for the exercise of normative powers.

These remarks indicate that the democratic peace generalization depends on a set of historically specific institutional and normative presuppositions having to do with states as the primary sources of organized political violence. When war is no longer the sole form of political violence, then the significance of the internal democracy of states as a means toward peace is greatly diminished. This is particularly true of the Kantian normative inference that democracies would somehow assure that the political federation of peaceful states is ever expanding. But once the institutional mechanisms of war making shift from representative bodies toward much less accountable administrative and executive functions and thus undermine the balance of institutional powers within a democracy, the expansive effect created by democratically organized institutions of domestic politics is less likely. This occurs when security requires limitations of freedoms and entitlements of one's own citizens.

Beyond these internal effects, security brings to a halt the expansion of the zone of peace among liberal democracies. This means that the borders of the zone of peace will become a source of political conflict with those who are outside it. By this I mean that various transnational publics are now increasingly aware of the "problematic fact" of the zone of liberal peace and prosperity and regard it as having inherent and systematic asymmetries. The increased potential for violence from those who are outside the zone of peace requires that democratic states adapt to these new threats to their security, often by restricting the liberties of their citizens and their own commitments to human rights, and thus leads to a tendency for democracies to restrict their own democracy and political inclusion within their own states as a result of threats to their security. In this way, the conditions and institutions that promoted a democratic peace among states now act as part of a new negative feedback mechanism, affecting particularly the liberties and rights that have permitted an active citizenry to possess enormous influence over the use of violence. Instead of democracies making international relations among states more peaceable, the new constellation of political violence is potentially making democratic states less democratic and less open to applying their internal standards of human rights and legal due process to those that they deem to be threats to security. Recent events show then that democratic peace depends on a positive feedback relation between the internal structure of states and the international political system, where democracy is internally promoted by external peace and external peace is promoted by wider powers of citizenship, including transnational citizenship. When citizenship is transnational, citizens can appeal directly to other institutions and associations in order to make states accountable, as is already the case with human rights violations. This mechanism has not been able to counteract the new negative feedback from the international system on democracy, and the negative and interactive effects of the emergence of the actual zone of peace indicate that its continued existence no longer depends solely upon the increased democratization of states. The fact that democracies do not wage wars against other democracies now means that the borders of conflict are externalized, by means that exact costs to their internal democratic character. The republican linkage between an empowered citizenry and international peace is in fact systematically severed.

If the practical import of these new feedback relationships undermines the prospect of expanding peace through a political union of existing democracies, peace and security are no longer reducible to the absence of war. Here we need to modify some deep assumptions about the proper location for democracy and the exercise of the powers of citizenship, in order to determine what would help democratic states to avoid the problem of the weakening of internal democracy as a means to maintain security. One possibility is that some supranational institutions could exist that would make democratic states more rather than less democratic. In a word, peace requires not democracies, but democratization at positively interacting levels.

These generalizations about the absence of war and famine are generally at the structural level, in that democratic institutions create conditions that make it difficult for political power to be exercised in such dominating ways. But this is not the only difficulty for democratic self-rule. It could very well be that even if democratic self-rule were robustly realized in the basic structure, citizens would be unable to take advantage of these opportunities due to failures of rationality. If we are to improve democracy, it is important not to deny such tendencies, but rather use social science to see the ways in which practical steps can be taken to create conditions under which such tendencies cannot operate unreflectively. Here a pragmatic approach inspired by Dewey could seek to improve human reasoning and deliberation directly, by discovering the contexts in which it not only fails, but also those in which it works well.

Skeptical generalizations about democracy: Improving human reasoning

As should be apparent from his moral psychology, Dewey argued that democratic aspirations to rationality and sound judgment were not chimerical, whatever the tendencies of habits to become rigid and influence thought below the level of conscious awareness. As any program of reform, his pragmatism sought to identify the appropriate circumstances under which such capacities worked well and could be further improved under the right conditions and education. This means that the aim of his social and moral psychology was to achieve the improvement of this capacity for all citizens, to bring about such conditions in which these possibilities are realized and the limitations of human thought are overcome. Thus, the challenge that Dewey faced already in his own time in the social psychology of Walter Lippmann and others is twofold. First, Lippmann and some contemporary social psychologists claim that the inherent tendency of human reasoning not only to be mistaken, but also so systematically mistaken as to undermine epistemic populism or epistemic egalitarianism. Second, if tendencies to violate ideal rationality are universal and innate, then the practical hope of improving human judgment and of eliminating biases is chimerical. Although never put in directly political terms, such claims are commonplace among naturalized social psychology of human reasoning in the "biases and heuristics" research program (Lee 2007).

If many empirical approaches to human behavior simply assumed basic human rationality in a variety of ways, Kahneman and Tversky saw their results of twelve different biases in human judgment as challenging "the descriptive adequacy of rational models of judgment and decision-making" (Kahneman and Tversky 1982, p. 494). This kind of research program required a particular methodological focus, "on errors and the role of judgment biases," often with regard to statistical and logical reasoning. While often cautious in their conclusions, Fischoff, Lee, and others point out that Kahneman and Tversky often are willing to overgeneralize their claims beyond the scope of their particular experiments, which others have argued often encourage the occurrence of errors in the subjects (Lee 2007, pp. 58–59). The important methodological counterattack to such experimental claims challenges the supposed the ubiquity of biases on the basis of such purely experimental data; if they are not overgeneralized, then they are highly context specific and do not show an inherent or general irrationality. In a similar vein, one could also inquire into "the conditions under which heuristics are valid" (Fischhoff 1982, p. 423). In such cases, the experimenter engages in what Fischhoff calls, using an engineering metaphor, "destructive testing," to see if there are conditions under which "a bias fails" with "the result of improving cognition" (Ibid.). The alternative to this more practical approach is to argue that the only way to improve inductive inferences is by giving people "inferential guides" to formal reasoning so that they are able to apply theoretical understanding of statistics and probability to everyday life (see Nisbett et al. 1982). Such a reliance on theoretical knowledge seeks as close an approximation to classical rationality as possible.

Whatever the force of its experimental evidence, this agent-centered research overlooks practical implications of those contexts and conditions that promote rational rather than irrational decision-making. Thus, experimental cognitive psychology could take a pragmatist turn and attempt to promote and facilitate rational and nonbiased judgment, as well as show the ways in which judgments under less-than-perfect conditions can be "debiased" and improved. In contexts in which the bias fails to influence inferences, the result is improved judgment, making such debiasing efforts highly practically significant. If rationality is conditional upon features of the context, then so is irrationality. Seen as dependent on both objective and subjective conditions, experimental results can guide improvements in judgments and deliberative procedures. One such procedure to lessen the effects of bias without undermining democracy is to see it functioning in a broader temporal context. Thus, opportunities to test could be distributed temporally across institutions, so that one type of deliberative body could test the decisions of another type of deliberative body.

Psychologists who engage in research with this ameliorative and practical aim call their normative conception of social research "applied cognitive psychology" (or ACP). ACP is a program consistent with the tenets of nonreductive naturalism insofar as it seeks not only debiasing, but also the practical goal of promoting norms of rationality in various decision-making procedures.

The key to such research is context sensitivity, a feature that Dewey identified as lacking in moral theories and necessary for all attempts to improve moral judgment. In the case of the conjunction fallacy that Kahneman and Tversky found to be common in the reasoning of the "Linda problem," Gigerenzer found that is possible for people to conform to the conjunction axiom, and the task is "debiased" and performance greatly improved if the probabilities were expressed as frequencies. (See Goldstein and Gigerenzer 2002, p. 75.) In Dewey's terms, frequencies better spell out the context in such a way that people can see what is at stake, what the situation is and what in it is problematic. Thus, Dewey's contextualism takes people to be reasoners who determine standards of correctness within particular situations, with particular problems and solutions. Practical moral thinking and judgment is always in interaction with a specific situation and social environment, and this ecological constraint should also extend to experimental methods as well.

One could formulate this response in more Deweyean terms by affirming that the social context of judgment provides an unavoidable practical constraint on understanding human judgment. Without some sense of the problematic situation in which judgment is operating, it is hard to see responses as "rational" and "intelligent" and thus as predictive of anything like a permanent tendency of human reasoning within an extended social environment. As an experiment with the purpose of uncovering internal features and tendencies of the agent, it is entirely unclear, when the contextualist constraint is honored, that there is in fact a "massive failure of the conjunction rule." At the same time, Tversky and Kahneman admit that the experiment was constructed so as "to elicit conjunction errors, and as such does not provide an unbiased estimate of the prevalence of these errors" (Kahneman and Tversky 1996).[7] Gigerenzer and his colleagues conducted experiments that attempted to assess whether or not there are central conditions that debiased judgments of the sort discussed by Kahneman and Tversky. Their results show the way in which a methodology sensitive to context specificity can demonstrate the way in which "we arrive at rational or irrational judgments in specific contexts of reasoning," including context in which subjects do the tasks particularly well.[8]

Such research sees a fallacy at work in taking various judgments to be inherently biased; to be inherently biased can only mean that there is some failure of rationality inherent to the human cognizer. This research shows, to the contrary, that improving judgments is not necessarily a matter of modifying such inherent features of human nature (as Dewey might put it), but of understanding the sensitivity to judgments to various environments and modifying the environment of choice and judgment so that people are able to reason better. Thus, an awareness of context sensitivity is a necessary requirement for improving human judgments on important moral, social, and legal matters, and proponents of Applied Cognitive Science think that a practically oriented study of human judgment could improve it by focusing on different conditions and environments as well as on possibilities of various debiasing techniques for more recalcitrant problems. Of course, such a study raises the same progressive questions as Dewey's moral psychology of the status of the social scientist in proposing standards, defining the limit of various heuristics, and constructing particular techniques of institutions that improve human reasoning. ACP provides a contemporary analogy to the type of interactive and normative naturalism that is central to Dewey's theory and his response to similar arguments for the cognitive limitations and bias of ordinary citizens among democratic realists such as Walter Lippmann. Dewey too sought to move away from global appraisal of human rationality to understanding specific cognitive processes in order to guide action. ACP studies normative phenomena nonreductively (that is, as normative phenomena), while at the same time offering evaluations of their rationality that are meant to also permit agents to improve them. For example, various debiasing techniques have been used practically for overcoming the self-serving bias, a major cause of impasses in negotiations. (See Babock et al. 2006.) This sort of ameliorative psychological and social science is still naturalistic, in the same sense that Dewey demanded: that such normative evaluations and proposals are themselves subject to experimental testing and contextual evaluation. As an achievement term, rationality broadens rather than narrows one life with the constant potential problem of the need for integration, often by transforming habits and accommodating impulses into an organization of various competing dispositions. In a way consistent with contemporary ideas of "ecological rationality," Dewey's improvements in human judgment more often than not mean improving the social conditions or environment under which decisions are made.

This sort of ecological rationality might prove useful with regard to several issues of moral and political psychology that Dewey confronted. Lippmann charged that the democratic ideal espoused by Dewey and other participatory democrats was too demanding and required "omni-competent citizens." Omni-competent citizens would be fully informed reasoners; in a word, they would be classically rational and equipped with a store of expert knowledge across many domains. As opposed to this requirement of effective participation, consider the "recognition heuristic" as an example of ecological rationality. When making inferences using it, people see recognition as correlating with a particular criterion, say the recognition of the name of a city with large population size. One might want to know which universities have the highest endowments and use a "mediator" such as a newspaper. Similarly, as Arthur Lupia shows, voters may correctly use party affiliation as providing just such an ecological correlation to various political beliefs, in order to decide whom to vote for in the absence of further information.[9] Both ecological correlations prove to be highly accurate, and in these cases, social institutions structure the environment in such a way that recognition is highly correlated to the appropriate criterion. Thus, changing or controlling the environment makes people smarter without necessarily promoting various moral and political virtues, however desirable these may be. Skeptics such as Lippmann are answered not by appeal to some inherent or populist rationality, but by facts about how institutions can be structured so as to promote environments for good, but frugal decision-making. This account of rationality within a context suggests that it is possible to mount a defense of the claim that ordinary deliberation may accomplish these ends, particularly when linked to inquiry where the environment is not informationally responsive. Such a defense would primarily be instrumental, that the deliberation of all those affected by a decision would be superior to any other possible method of inquiry. An interactive or ecological approach to such issues already provides some defense of such a claim, to the extent that it says that such a procedure would be the best only if certain social or environmental conditions are met. Creating such conditions is the task of democratic reform.

Conclusion

Both Critical Theory and pragmatism suggest that the social scientific study of democracy becomes one aspect of a practical theory or praxeology oriented to improving democratic practice. The central questions for a practically oriented social science of democracy are the following: What available forms of *praxis* are able to promote the transformations that could lead to new forms of democracy? What sort of practical knowledge is needed to make this possible, and how might this knowledge be stabilized in institutionalized forms of democratic inquiry? What are the institutional means available to improve democratic deliberation, especially given the fact that certain contexts clearly promote worse rather than better reasoning and deliberation? One very robust finding is that deliberation within heterogeneous groups is less susceptible to framing effects that deliberation in homogeneous groups. Such a finding has clear practical import for deliberative practice.

The forms of social science and types of social scientific reasoning discussed here are quite diverse. Together they provide a cumulative argument that there is no privileged form of social science that provides the basis for practical normative inquiry of this sort. The democratic ideal has many different features, and the circumstances in which it is realized are quite various. In this respect, democracy institutionalizes the proper methods of reasoning along with institutions that create the appropriate environment for testing and revision. It does so without stipulating what the proper standards or aims ought to be in advance. As a form of inquiry, social science can participate in this collective process, but not by claiming expertise about democracy itself. If the form of democracy is deliberative, then it too is a form of inquiry of citizens reasoning together cooperatively. Given that these institutions also stand in relations of reciprocal determining with changing circumstances, the norms that inform democratic institutions are always defeasible, awaiting improvement through further inquiry.

Notes

1. For a related criticism, see Zolo 1997.
2. On the various positions in the controversies over globalization, see Held, McGrew, Goldblatt, and Perraton 1999.
3. See Held 1995, pp. 98–101, for an argument of this sort that emphasizes the scope of interconnections and its consequences for the realization of autonomy as the key problem for democratic governments.
4. For various dimensions of this issue, see Hurrell and Woods 1999.
5. Here I am rejecting Held's earlier formulation of how the norms of democracy ought to be rethought in light of globalization. See Held 1995, p. 145.
6. As Russett puts it: "Depending on the precise criteria, only twelve to fifteen states qualified as democracies at the end of the nineteenth century. The empirical significance of the rarity of war between democracies really only emerges in the first half of the twentieth century, with at least twice the number of democracies as earlier, and especially with the existence of perhaps sixty democracies by the mid-1980s." See Russett 1993, p. 20.
7. On these methodological problems, see Lee 2007, p. 61ff.
8. For a discussion of the origin of Applied Cognitive Psychology as a practical and ameliorative research program, see Lee 2007, pp. 62–63.
9. For a discussion of the fruitfulness of such a heuristic, see Lupia and McCubbins 1998.

References

Agarwal, Bina (2001) 'Conceptualizing Environmental Collective Action: Why Gender Matters.' *Cambridge Journal of Economics* 24, 283–310.

Babock, L., J. Lowenstein and S. Issacharoff (2006) 'Creating Convergence.' *Law and Social Inquiry* 22, 913–25.

Bohman, J. (1999) 'Theories, Practices, and Pluralism: A Pragmatic Interpretation of Critical Social Science.' *Philosophy of the Social Sciences,* 28, 459–80.

Bohman, J. (2002) 'Critical Theory as Practical Knowledge.' In: *Blackwell Companion to the Philosophy of the Social Sciences,* ed. P. Roth and S. Turner. London: Blackwell.

Dahl, R. (1989) *Democracy and Its Critics.* New Haven, CT: Yale University Press.

Dewey, J. (1986) 'Logic: The Theory of Inquiry.' In: *The Later Works, 1938,* vol. 12. Carbondale: Southern Illinois University Press.

Dewey, J. (1988) 'The Public and Its Problems.' In: *The Later Works, 1925–1927,* vol. 2. Carbondale: Southern Illinois University Press.

Dewey, J. (1991) 'Liberalism and Social Action.' In: *The Later Works, 1935–1937,* vol. 11. Carbondale: Southern Illinois University Press.

Fischhoff, B. (1982) 'Debiasing.' In: *Judgment under Uncertainty,* ed. D. Kahneman, P. Slovic, and A. Tversky. Cambridge: Cambridge University Press.

Giddens, A. (1990) *Consequences of Modernity.* Stanford, CA: Stanford University Press.

Goldstein, M. and G. Gigerenzer (2002) 'Models of Ecological Rationality: The Recognition Heuristic.' *Psychological Review* 100, 75–90.

Habermas, J. (1974) *Legitimation Crisis*. Boston: Beacon Press.

Habermas, J. (1996) *Between Facts and Norms*. Cambridge, MA: MIT Press.

Habermas, J. (2001) *The Postnational Constellation*. Cambridge, MA: MIT Press.

Held, D., A. McGrew, D. Goldblatt and J. Perraton (1999) *Global Transformations: Politics, Economics, and Culture*. Stanford, CA: Stanford University Press.

Held, D. (1995) *Democracy and the Global Order*. Stanford, CA: Stanford University Press.

Hollis, M. (1977) *Models of Man*. Cambridge: Cambridge University Press.

Hurrell, A. and N. Woods, eds (1999) *Inequality, Globalization, and World Politics*. Oxford: Oxford University Press.

Kahneman, D. and A. Tversky (1982) 'On the Study of Statistical Intuitions.' In: *Judgment under Uncertainty*, ed. D. Kahneman, P. Slovic and A. Tversky. Cambridge: Cambridge University Press.

Kahneman, D. and A. Tversky (1996) 'On the Reality of Cognitive Illusions.' *Psychological Review* 103, 582–91.

Kant, I. (1970) *Political Writings*, ed. Hans Reiss. Cambridge: Cambridge University Press.

Lee, C. (2007) 'Applied Cognitive Psychology and the Strong Replacement of Epistemology by Normative Psychology.' *Philosophy of the Social Sciences* 38, 55–75.

Linklater, A. (2001) 'The Changing Contours of Critical International Relations Theory.' In: *Critical Theory and World Politics*, ed. Richard W. Jones. London: Lynne Rienner.

Lupia A. and M. D. McCubbins (1998) *The Democratic Dilemma: Can Citizens Learn What They Need to Know?* Cambridge: Cambridge University Press.

Nisbett, R., D. Krantz, C. Jepson and G. Fong (1982) 'Improving Inductive Inference.' In: *Judgment Under Uncertainty* ed. D. Kahneman, P. Slovic and A. Tversky. Cambridge: Cambridge University Press.

Rawls, J. (1999) 'The Idea of an Overlapping Consensus.' In: *Collected Papers*, ed. Samuel Freeman. Cambridge, MA: Harvard University Press.

Rodrik, D. (1994) *Has Globalization Gone Too Far?* Washington, DC: Foreign Affairs Press.

Russett, B. (1993) *Grasping the Democratic Peace*. Princeton, NJ: Princeton University Press.

Sen, A. (1986) *Poverty and Famine*. Oxford: Oxford University Press.

Sen, A. (1999) *Development as Freedom*. New York: Knopf.

Tilly, C. (1990) *Coercion, Capital and European States*. London: Blackwell.

Toulmin, S. (1990) *Cosmopolis: The Hidden Agenda of Modernity*. Chicago: University of Chicago Press.

Wimsatt, W. (1974) 'Complexity and Organization.' In: *Proceedings of the Philosophy of Science Association 1972*, ed. R. S. Cohen. Dordrecht, The Netherlands: Riedel, 67–86.

Woods, N. (2001) 'Making the IMF and the World Bank More Accountable.' *International Affairs* 77(1), 83–100.

Zolo, D. (1997) *Cosmopolis: Prospects for World Government*. Cambridge: Polity Press.

5
Fact and Value in Democratic Theory
Harold Kincaid

There has been a long and interesting history of interaction between normative democratic theory, the social sciences, and the philosophy of science. Defenders of liberal political philosophy have often been the same individuals who study how democracy works; much social scientific study of democracy has been motivated by the normative beliefs that democracy has a variety of virtues, and philosophers of science have both been public advocates of liberal democracy and seen parallels to it in the proper normative values of good science. My goal is to explore some of these interactions.

My main thesis is that certain aspects of the social science studies of democracy, of democratic political theory, and of philosophy of science share a common core set of assumptions that are mistaken. Their projects assume that practices can be explained and evaluated by identifying the formal procedures that are followed, that those procedures can be identified and understood independently of their social embodiment, and that they can be so understood in a value neutral way.

My discussion breaks into three parts: I consider first the social science assumptions of democratic normative theory, next the value assumptions of social scientific studies of democracy, and finally the mix of both social science assumptions and democratically inspired epistemic norms in the philosophy of science.

Some background

It will be useful first to outline some general ideas about the role of values in science and about differing conceptions of social processes that someone might adopt in thinking about the values of democracy and the sociological understanding of democracy and of science.

I take for granted the Quinean idea of the web of belief—that our beliefs are connected in multiple and cross-cutting ways. This does not mean that every belief is connected to every other one in some kind of seamless whole, but it does mean that we should expect to find interesting, diverse, and complex interconnections among our beliefs. I also take for granted that recent science studies have shown that much science is an inherently social process (which is not to say it is a mere construction). Given these assumptions, we should expect there to be a variety of complex interactions between fact and value in both democratic normative theory and in the social science explaining democracy. In particular, when asking about the role of values in science, we need to distinguish the following logically independent issues (Kincaid, Dupre, and Wylie 2007):

- The kinds of values involved
- How values are involved
- Where values are involved
- What effect their involvement has

Traditionally, epistemic values are sharply separated from moral, political, and pragmatic values of various sorts. I just note that, in practice, the distinction is not always so clear.

Values might be involved in very different parts of science, and it would be a mistake to ignore these different sources. Values might be involved in the questions we ask and the topics we pursue, and they might be involved in the many different uses to which scientific results are put, both scientific and intellectual. That is very different than saying that they are involved in the heart of science—in providing evidence and explanation.

To say values are "involved" is also very vague, for they might be "involved" in very different ways. Here are some possibilities. Values might be involved essentially or accidentally. The science in question may be logically committed to certain values or the values might be more or less tangential. Epistemic values are a prime example of the former and blatant bias an instance of the latter. Moreover, values might be involved directly—they may be presupposed as it were—or by conversational implicature or by implication. I would argue that this is an important distinction, and that involvement by implicature is an important phenomenon in understanding science. Scientific results and texts are not just free-standing platonic entities—they are embedded in a variety of social and intellectual contexts where they can have implications that go considerably beyond their surface content.

Finally, a crucial question concerns what effect you take the presence of values to have. The traditional claim is that nonepistemic values preclude objectivity and epistemic values promote it. That assumes a blanket subjectivism about moral values that I would reject. But even granting that I do not think the presence of values inevitably implies subjectivity and bias—the values may be explicit and shared. And there are certainly occasions in science where epistemic values are in doubt or conflict, raising questions about objectivity as well.

Turning from values to conceptions of the social brought to the study of democracy and science, it is worth noting that the social can be conceived either in thick or thin terms. The inspiration here is Geertz's (1973) distinction between thick and thin interpretation and Williams's (1985) thick and thin moral concepts. Social factors can vary along two dimensions that give them a thicker or thinner place in our explanations. The first concerns their relative causal importance, where that might be measured either by size or by modality. Size of a cause might be construed by how much effect wiggling one cause has compared to another. Modality might concern whether a cause is necessary or sufficient or whether it is trivially necessary (as the sun is for any social explanation) or importantly necessary (as evolutionary past may be for the explanation of social phenomena).

Thick or thin social explanations may also be distinguished by the extent to which explanations can be given in terms of individuals described in nonsocial terms. Rational choice explanations that only require knowledge of utility functions that do not include other individuals are as about as thin as you can get. Sociological explanations in terms of macrosociological entities such as classes are at the opposite thick end.

Of course, the above notes on values, science, and the social are abstract and certainly could be fleshed out in more detail. However, they suffice for my purposes and will get some more content as I make my arguments about science, values, and democracy.

The social assumptions of normative theory

I argue in this section that much current normative work in democratic theory is not well supported by solid findings in the social sciences or, more strongly, that there are solid findings in the social sciences that are inconsistent with the assumptions of normative democratic theory. One basic problem comes from an implausibly thin notion of the social in the form of treating democratic processes as separable from factors in larger societal organization. Another comes from other diverse instances of getting the social facts wrong.

So I take it that there is good social science evidence that (Domhoff 2005):

- In Western democracies there exists a small elite that has control of enormous amounts of material wealth.
- That elite has substantial influence on most aspects of society through control of corporations, interlocking directorships of the same, and through their presence on the boards of foundations, universities, and other public institutions.
- In pursuing their interests, the small elite that controls the corporations and other institutions have significant influence on public policy through

campaign contributions, bribes, control of media, filling public offices, and so on.

These theses are consistent with but not equivalent to the Marxist claim that the state is an instrument of the ruling class. It is not equivalent because it leaves plenty of room for self-serving bureaucracies, for interest groups other than the wealthy elites, and for elites that may not constitute classes in any strong sense.

Some other relevant results include

- State organizations develop interests of their own—bureaucracies protect their turf and so on.
- Legislation and policy implementation are substantially influenced by legislators and other government officials pursuing their own interests in interaction with organized interests groups doing the same.
- Ethnic groups and more local communities likewise have elites with disproportionate access to resources and disproportionate influence. In short, local communities are not homogeneous happy families—there are important competing interest groups.
- The modern democratic state reached its current form after extensive battles with various excluded groups.

All the above point to the fact that democratic political processes are strongly embedded and influenced by the surrounding social environment and cannot be fully understood without identifying its role.

I want next to argue that contemporary normative democratic theory is in deep conflict with these empirical facts along with others that are not so neatly subsumed under this rubric and that I will bring up as I go along. I focus first on one idea common to much democratic theory—the notion of the common good—and then on the specific ideas of four contemporary normative theorists: Dahl, Kymlicka, Sandel, and Pettit.

The idea of "the common good" is repeatedly used in contemporary democratic political theory, especially by those in the communitarian tradition, to justify liberal democracies or to talk about how they might be improved. Yet it is very hard to find a clear statement of what the common good is. Probably the most rigorous notion of the common good is the economists' notion of a Pareto improvement in which a proposed change makes at least some better off and none worse off. Utilitarian readings are possible—the common good is the greatest good—but they are often not consistent with the political philosophies being advocated. In any case, the social science results described above make it unlikely that there is any very useful notion.

Given the strong division of interests between groups in contemporary society, there are unlikely to be many policies, practices, and so on that favor the interests of some while not demoting the interests of others. Finite resources have to be divided, and what goes to one group is subtracted from what others can have. Similarly, given that power relationships are ubiquitous throughout contemporary societies, improving the situation of one group tends naturally to be at the expense of other groups, because improving the situation of a group often increases its ability to compete. Furthermore, even if we have a policy or practice that allegedly makes every one better off, we still have to answer the following questions: With respect to what other possibilities? And while holding what fixed? If there are deep divisions of interests between groups in society, then how we specify those possibilities and fixed conditions will matter, for different possibilities will have different effects for different individuals and groups.

I turn next to look at questionable empirical social science assumptions in the work of specific contemporary normative political theorists. In his recent *On Democracy* (2000), Dahl makes the following claims:

• Representative democracy exists because of size factors—it is the only way for large groups of people to govern themselves effectively.
• Democracy ensures personal freedom that extends to our ordinary life.
• In a democracy, we live under laws of our own choosing.
• Democracy is a good system because it furthers human development.
• In a democracy, there is full control of the military and police.

You might think that Dahl's account here is a relativized one to the effect that liberal democracies do this better than any existing alternatives. However, he clearly does not make that qualification, with the implied conclusion that they do it well.

Dahl's vision of democracy is shared by those from rather different traditions in democratic theory, namely, those from roughly the communitarian tradition. Kymlicka (2001) has written a great deal in this tradition, focusing on minority rights (he is Canadian and concerned about the status of the French-speaking population). Kymlicka emphasizes the virtues of citizens that he thinks are necessary for a democracy to be stable and function effectively. Chief among these is the need to give public reasons—not just statements of preference, but arguments that should have force for everyone. When this and other related virtues of citizens are present, liberal democracies are able to promote the general welfare, effectively manage the economy, and so on.

Sandel (1998) shares Kymlicka's concern with the virtues of citizens necessary for democracy to do well. Democracy provides self-governance only if citizens feel a moral bond with others in their relevant communities. Self-governance thus requires that the government cultivate the traits in individuals that make this possible. When it does so, individuals participate in communities that "control their own fate." Sandel thinks that this republican ideal predominated in American democratic politics prior to World War II, but has faltered since.

The republican ideal is defended at length by Pettit (1997). Key to that ideal is the idea of nondomination. Nondomination is compatible with restricted choices and is not ensured by traditional liberal noninterference according to Pettit. It is a more worthy ideal than that expressed by traditional liberalism through notions of negative liberty and noninterference. Of interest to me is what Pettit says about the state, democracy, and nondomination. On his view, nondomination is something "the state is able to pursue fairly effectively" (p. 92). Moreover, it is not just that the state can causally effect this, but the "institutions [of the state] will constitute... the very nondomination which citizens enjoy under them" (p. 107). Nondomination comes about when institutions of the state make their decisions "contestable" and the institutions of the modern state can realize this virtue through the ability to write your elected representative, to appeal judicial decisions, and through rights of association and expression (p. 193).

As far as I can see, all four of these political philosophies are at odds with the social science results mentioned above. Contra Dahl, democracy exists because the nation state outcompeted other forms and then was forced to the current system by the social movements rather than being an optimal solution to the problem of how a group of individuals could agree to self-governing institutions. Here the functionalist assumption is in principle an objective, empirical matter, but it also has value implications, for example, democracy is optimal.

If you keep the actual history in mind rather than Dahl's functionalist optimism, then other components he attributes to democracy also come into question. So the military and police are not under popular control but are instead organizations pursuing their own interests successfully with some constraint by the larger society. Similarly, the "personal freedom in ordinary life" that Dahl sees in democracy is denied to many who have to spend 50 hours a week in mind-numbing jobs where their every moment is under supervision and control and the result is to barely eke out an existence. This is the kind of point that a defender of Republicanism like Pettit can make in criticizing liberalism.

Contra Sandel and Kymlicka, I think the evidence is clear that citizens do not control their own fate because of the enormous influence of ruling elites and all the other manifestations of power and interest in political decision-making. Of course, this is always a matter of degree. Yet Sandel and Kymlicka seem to think that current forms of democracy can well realize their ideals. So Sandel thinks that the United States prior to World War II was a place of "moral bonds" where citizens did control their own fate. The truth is, that it was a place where there was racial apartheid, rampant gender discrimination, and so on.

Finally, contra Pettit, the kinds of elites, interest groups, bureaucracies with vested interests, and other groups that I pointed to above make his claims about how the state can function look rather optimistic. He does note that contestation and thus nondomination are most likely when there are social movements, something the social science research bears out I believe. However, successful social movements are rare, and most every day democratic processes go on untouched by them.

The discrepancy between reality as described by good social science work and that presumed by democratic theorists is so great that it calls for an explanation. This is where I think there is an interesting possible use for the social sciences relevant to political theory. I have in mind the sociology of knowledge question concerning how democratic political philosophies are developed, interpreted, and spread. This suggests that I do not have in mind explanations such as "intuiting moral facts" or "went where the arguments lead them," although this is not to deny that arguments are given or deny that deep moral intuitions have a role. Nor do I have in mind a crude "the ruling ideas are those of the ruling class" account. Rather, I am asking about the sociological factors such as sources of funding and determinants of differential uptake of ideas that determine which political philosophy gets attention. There is more to the story than that there are philosophers who work for universities and students who do or do not pass on their ideas.

Values and the study of democracy

I want to turn next to look at the connections running the other way—at the value presuppositions behind social scientific studies of democracy. I focus on studies of democratization—the process and causes of extending and deepening democracy. There have been many studies on the causes of democracy and of the extent of democracy: is it economic growth, political culture, literacy, or something else? Of course, to study democracy, you have to know what it is, and how it is separated from other governmental forms and how a more developed democracy is distinguished from a less developed one. Perhaps not surprisingly, defining democracy raises normative issues in several ways. The first is in the construction of measurement indexes. These indexes are put through extensive statistical testing of various sorts to ensure both the reliability of the measures—that different raters produce the same results—and their validity, that is, that the measures pick out antecedently obvious cases of democracies. Nonetheless, I would argue that value judgments are involved in several ways in the process.

Here is one example. Bollen (1993) has produced a measure of democracy that is widely used in empirical studies. His measure requires that elections be "fair." What does "fair" come to? For Bollen, fairness requires that elections are "relatively free from corruption and coercion," which in turn depends on whether "alternative choices exist," "are administered by a non-partisan administration," and are not "rigged."

Not exactly paradigm descriptive terms, are they? Values are involved here in multiple ways. More obvious perhaps are the inevitable values involved in determining which elections are fair. This comes out quite clearly when we note that Bollen's measure of democracy puts the United States at the highest reaches of the scale in 1960. What was U.S. democracy like in 1960? In 1960 in the United States, apartheid reigned in the South—housing, employment, and public accommodations were strictly segregated, and the system was enforced by explicit repression when necessary. African Americans were effectively prevented from voting in many states. Also, elections were frequently run by the party in control. A famous instance in 1960 was the control of the Chicago ballots by Mayor Richard Daley, later famous for the police riots at the 1968 Democratic convention. The Chicago political machine waited until all other ballots from Illinois were in so that they could produce the correct number in Chicago and put Illinois in the Kennedy camp, which itself determined who won the national election.

So treating the United States at this time as the epitome of democracy is surely a controversial normative judgment, especially when we see that Bollen counts the United States as more democratic than other European democracies at the same time. This is pretty common in democracy studies, at least by Americans; it is taken as a given that the United States is the epitome of democracy and everything else is judged in that light. Values are also involved in the judgment that there is a choice in elections. For example, there were arguably very few major policy issues where Nixon and Kennedy differed. Of course, some might argue that there were indeed major differences, but it is hard to believe that such arguments or criticisms of them can be made independently of prior normative assumptions.

What effect do these value assumptions have on the objectivity of the results? They are all assumptions that arise in the processes of providing evidence, so they are not peripheral to the science. Yet they have different roles there with different consequences.

The values built into the measures of fair elections and the extent to which there is choice between parties can compromise objectivity. If I do not share your values in these cases, then I am not going to accept your empirical claims about which countries are more democratic and about the causes of democratization. On the other hand, if those value judgments were made explicit and fully shared by those who might disagree, then there could be some kind of objectivity, granted those assumptions. However, the values in this research usually are not entirely explicit, and it is not clear how widely they were shared by non-U.S. researchers.

Less obvious value assumptions arise because these definitions look for a procedural criterion for democracy—they take the essence of democracy to be the following of rules and procedures. But this is not the only way to think of democracy. Shapiro (2005), for example, argues persuasively on my view that the key virtues of democracy are providing some constraints on the illegitimate use of power and that purely formal measures of democracy do not necessarily capture this. That conclusion fits well with my claim earlier that thin notions of the social are often invoked when studying democracy and doing political theory, despite the fact social science research seems to show that democracy is much more heavily influenced by social organization. However, using measures that focus on procedures alone takes a stand on the normative issue of what justifies democracy.

This is an instance of value assumption by implication and has no direct consequences for the objectivity of the empirical work so far as I can see, again illustrating my earlier claim that values can be involved in different ways with different consequences in science. Studying democracy in procedural terms does not make value judgments that are essential to the objectivity of the evidence claims concerning the causes and effects of democracy picked out procedurally. Someone who disagrees with the procedural justification of democracy can nonetheless accept that these results are well confirmed for democracy conceived as meeting procedural criteria.

I want to next show how similar value assumptions are at work in studies of "new' democracies. Political scientists, particularly U.S.- based ones, have tried to make sense of the political systems in the territories of the former Soviet Union. They have come to categorize them as "imperfect" and "immature" democracies. They do so because those systems fail to realize the standard formal criteria—fair and free elections, for example—to varying degrees. But they are not dictatorships either. So there has come to be a substantial literature on "qualified democracies" attempting to analyze and explain how they work.

These studies are interesting grist for my mill again because they make normative assumptions. They ignore the extent to which there is a ruling elite of wealth and power no matter how democratic the best examples may be, and they tend to emphasize the importance of democratic trappings over substantial issues about the extent to which there are limits to the exercise of power. So there are competitive authoritarian regimes—regimes on which there are no or minimal democratic trappings but nonetheless interest groups with sufficient power to put limits on what the state can do—that may well fare better on criteria such as Shapiro's than the imperfect democracies. So value implications come when they prioritize the formal over the substantive—they are favoring one normative theory over another. As I argued above, this sense of value-ladenness may not be an obstacle to objectivity in itself. However, there are still other worries, namely, that most of the practices that are cited as indicative of "immature" democracy arguably have been and/or are still being practiced in the United States. Former president George W. Bush's various successful attempts to avoid legislative and judicial control are prime examples. Any empirical comparison that tries to explain the differences between mature and immature democracies is, to that extent, biased in its recording of the evidence in that nations are assigned to categories in a way that is not supported once the U.S. bias is removed.

Some morals for democratic science and philosophy of science

I turn finally to look at some interactions between philosophy of science, democratic theory, and social science. A basic idea behind liberal political philosophy is that democracies produce outcomes that result from the free competition of ideas. Also part of that tradition is the thought that outcomes that result from voluntary decisions of rational self-interested individuals are ones in effect agreed to by all. These powerful ideas have deep reflection in the philosophy of science up until the present, both in normative assessment of what constitutes good science and in descriptive accounts of how science works.

It is an interesting historical question examining how and where these democratic ideals have interacted with philosophy of science in the last decade, but interaction there has been. An interesting recent example is Ian Jarvie's (2001) work on Popper. Jarvie argues fairly convincingly that there is an ignored social strain in Popper's thought that Jarvie labels the "Republic of Science." The free market of ideas concept seems at work here, what we should expect from the author of *The Open Society and Its Enemies*.

I have argued in previous sections that normative democratic theory often rests on dubious social science—overly thin notions of the social— and that social scientific study of democracy not only does that but also makes important normative assumptions in the process. In this section, I argue that all of these mistakes show up in naturalized philosophy of science inspired by the concept of the free market of ideas.

More recently, the influence of the marketplace of ideas concept on philosophy of science shows up clearly in the work of Kitcher (1993) and Zamora (2002). Both use rational choice and/or game theory to analyze science in a way that either implicitly or explicitly invokes the market of ideas side of democratic theory. Kitcher provides models where optimizing self-interested agents with limited commitment to the truth can produce reliable results. Both authors are advocates of naturalism—a scientific study of science itself—and both see their accounts as contributing to that goal.

Zamora is the most explicit about the political economy foundations of his philosophy of science, and I want to focus on his models in asking how the social is brought into these naturalistic accounts. In a series of innovative papers using rational choice and rational choice game theory to analyze scientific practices, Zamora argues, among other things, that

- Recognition-seeking scientists will have a preference for some scientific norms over others.
- A community of scientists could reach agreement on a norm to determine who should be the winner in a race for recognition.
- It is "the free agreement of individuals" that determines what such norms are.
- These choices are made under a "veil of ignorance" in that scientists must consider that they will be in a variety of different scientific situations in the future.
- This veil of ignorance ensures that norms favoring unbiasedness and sincere reporting will be chosen.
- Norms in science are "compromises" given the interests and preferences of individual scientists.

So the moral for Zamora is that there is a social contract behind some of the methodological norms in science.

I fear that Zamora has provided us with an impoverished conception of social processes involved in science for several reasons. For starters, there is much science where free agreement of individuals behind a veil of ignorance looks like a fairly naïve naturalistic account. A real-world example that illustrates my concerns comes from the standards for acceptable evidence in medicine. There is no one clear criterion for how much evidence is enough across biomedical research. For much of the profession, a randomized clinical trial is essential and that is enshrined in the Federal Drug Administration (FDA) requirement for two confirmatory Phase III randomized clinical trials (RCTs). Power first shows its head in this context in terms of which RCTs count as evidence. Drug companies almost never report a negative result (they are buried) because it is not in their interests—the vast majority of drug company supported studies find a positive effect (Sismondo 2008). This effect is compounded by the fact that surveys show that one-third of biomedical researchers have commercial funding and that most research contracts often give the commercial funder say over what is published.

Another place power shows its face is in terms of what parts of the medical profession are subject to the RCT standard. Surgeons in the United States are generally not. New surgical procedures are devised and propagated without RCTs testing their validity. They have a vested interest in doing so—use of RCTs would limit the number of procedures they could do. Needless to say, surgeons are a well-organized group with considerable influence.

A second reason I fear that Kitcher and Zamora have an inappropriately thin account of the social comes from the deep limitations of the rational choice version of game theory they employ. Rational choice theory and rational choice game theory do not really have a place for social processes. The results of Kitcher and Zamora are static results, as is much of rational choice theory and rational choice game theory that analyzes equilibrium solutions without describing a mechanism that produces equilibrium. Even repeated games are not really dynamic—the dynamics are folded into hypothetical reasoning. This is troubling in part because it is not obvious what conclusion to draw from static results when the dynamics are not known. This is particularly true when attempts to model real mechanisms do not give credence to the static outcomes. I fear that prospect is likely for rational choice studies of scientific norms. Evolutionary game theory models that are dynamic do not favor the static results of game theory. Models incorporating learning in the form of imitation produce outcomes that can reach no equilibrium or equilibria far from what would be optimal. Since imitation is a well-known dissemination process in science, this seems to pose a real problem.

Thus, there are serious empirical obstacles to drawing conclusions from rational choice accounts of science because they have a much too thin conception of the social, just as in the democratic normative theory that partly inspires them. I now want to argue that there are also normative assumptions in these apparent naturalistic explanations of science. Here Kitcher is my focus.

The purpose of modeling science as the outcome of rational self-interested actors is to show that social processes in science could be compatible with a science that tracks the truth. Sullied individuals—those with impure motives—could nonetheless produce epistemically good results if the incentives are constructed correctly. Kitcher (1993) combines these models with a general philosophical argument that mature science must be on the right track because that is the best explanation for key features of science—basically the success of triangulation.

I would argue that there are foundationalist epistemic values working here despite Kitcher's avowed naturalism. An argument showing how science might work seems to me to carry weight only if there is already a presumption that modern science is on the right track in the first place. However, Kitcher's models are much too far from reality and real empirical studies of actual science to have any weight as explanations for how science does actually work. Rather, Kitcher's positive argument for realism is a philosophical one that makes no attempt to show empirically that social factors work in an epistemically favorable way. Kitcher is taking for granted that modern science can be evaluated more or less as a unit that has success and that philosophers can explain that success.

I would counterpose to these assumptions a more thoroughgoing naturalism that would deny that there is any automatic presumption that modern science is on the right track. Mature modern science refers to an enormous batch of practices, of theories variously interpreted, of piecemeal claims that are needed to tie theories to reality, and so on. Thus, it is not something susceptible to uniform analysis. I would also argue that arguments of the form "science has characteristic X and they could only have X if they were truth related" have a suspiciously transcendental look that gives philosophical arguments a status they should not have.

Like the social science research on democracy that I discussed earlier, these attempts at social science explanations of science make value assumptions (about epistemic values) in part because of their hope to focus on formal procedures. Kitcher and Zamora both argue that science is characterized by certain procedures, and because they have these procedures, they are on track to produce successful science. However, I doubt that the procedures of science can be understood on their own any more than can the procedures of democracy; for example, meeting the gold standard of the successful procedures of an RCT is no guarantee of warrant. It is interesting to recall here Reisch's (2005) work showing that positivists like Neurath thought that we needed a thick sociology of science to evaluate science, that these ideas were attacked and driven out of philosophy of science because they were thought to be associated with the political left and antithetical to the free market of ideas, and that philosophy of science emphasized primarily the formal side of science as result, vestiges of which are still found in writers like Kitcher and Zamora. Ideals of democracy, of social science, and of the philosophy of science are indeed intertwined.

My point is neither to deny that scientific norms can sometimes be the result of rational choice social contracts nor to deny that rational choice models are useful. Rather, the conclusion I draw is that like political normative theory and some variants of empirical democratic studies, fundamental sociological processes are left out and that questionable value assumptions—in this case epistemic ones—may be made in the process.

In conclusion, I hope to have shown that there is a complex interaction between empirical social science and democratic normative theory that is mirrored in the philosophy of science and that progress in all three areas may require developing a more sophisticated understanding of social processes.

References

Bollen, K. (1993) 'Liberal Democracy: Validity and Sources Biases in Cross-National Measures.' *American Journal of Political Science* 37, 1207–30.

Dahl, R. (2000) *On Democracy.* New Haven, CT: Yale University Press.

Domhoff, W. (2005) *Who Rules America?* New York: McGraw-Hill.

Geertz, C. (1973) 'Thick Description: Toward an Interpretive Theory of Culture.' In: *The Interpretation of Cultures: Selected Essays.* New York: Basic Books, pp. 3–30.

Jarvie, I. (2001) *The Republic of Science.* Amsterdam: Rodopi.

Kincaid, H., J. Dupre and A. Wylie (2007) *Value Free Science: Ideals and Illusions.* Oxford: Oxford University Press.

Kitcher, P. (1993) *The Advancement of Science.* Oxford: Oxford University Press.

Kymlicka, W. (2001) *Politics in the Vernacular.* Oxford: Oxford University Press.

Pettit, P. (1997) *Republicanism.* Oxford: Oxford University Press.

Reisch, G. (2005) *How the Cold War Transformed Philosophy of Science.* Cambridge: Cambridge University Press.

Sandel, M. (1998) *Democracy and Its Discontents.* Cambridge, MA: Belknap Press.

Shapiro, I. (2005) *The State of Democratic Theory.* Princeton, NJ: Princeton University Press.

Sismondo, S. (2008) 'Pharmaceutical Company Funding and Its Consequences: A Qualitative Systematic Review.' *Contemporary Clinical Trials* 29, 109–113

Williams, B. (1985) *Ethics and the Limits of Philosophy.* Cambridge, MA: Harvard University Press.

Zamora, J. (2002) 'Scientific Inference and the Pursuit of Fame: a Contractarian Approach.' *Philosophy of Science* 69, 300–323.

Part III

Democratic Theory Elucidating Social Scientific Theory and Practice

6
The Problem With(out) Consensus: The Scientific Consensus, Deliberative Democracy and Agonistic Pluralism

Jeroen Van Bouwel

Introduction: Inquiring the scientific consensus ideal

The emergence of a consensus theory replacing another theory, or set of theories, is often interpreted as a proof of scientific progress and a marker of truth; ideally, all scientific inquiry and debate would result in a consensus. This ideal motivates, for instance, consensus conferences aiming to craft a consensus on controversial scientific matters to inform the public or stakeholders (e.g., on best medical practice or on environmental issues). Finding a scientific consensus then proves the validity of a theory and – indirectly – of the public policy based on the consensus theory.

The other side of the coin is that the lack of scientific consensus often is used to undermine or criticize science and the public policy based on it (e.g., former U.S. President George W. Bush on climate change). When scientists agree, their results are taken more seriously than when they disagree, even though such an agreement or consensus might hinder scientific progress because of critical, heterodox theories not being taken seriously (e.g., the theory of continental drift was accepted by geologists only after 50 years of rejection, and the theory of Heliobacter pylori as the cause of stomach ulcers, was at first widely rejected by the medical community).

These observations might question scientific consensus as an ideal or as the goal of inquiry and marker of truth: Is there an epistemically more desirable ideal to be found? What are the epistemic benefits of the lack of consensus, of continued dissent? Can we elaborate a (meta)consensus on dissent and scientific pluralism? Tackling these questions in this chapter, I will explore the symmetries between *models of democracy* and *models of science* if it comes to elaborating consensus or cherishing dissent. This gives us the opportunity to compare interpretations of dissent and pluralism in science and democratic theory and to make the idea of scientific pluralism in the social sciences more precise. The significance of this exercise for the social sciences will be illustrated by some examples from sociology and economics.

Consensus and deliberative democracy

The most obvious candidate when looking for a democratic theory that can elucidate the idea(l) of consensus is deliberative democracy.[1] Much (but not all) of the deliberative democracy literature takes the rationally motivated consensus as the goal of deliberation. Two prominent figures in this literature are Jürgen Habermas and John Rawls; both of them, and their followers, rely on the idea of an unforced consensus.

Habermas and his disciples follow the idea that something can be justified if it results from universal unforced agreement in an ideal deliberative situation, the ideal speech situation – very much like a graduate university seminar – characterised by the consensus-generating force of arguments leading to the rational outcome. (Obviously, obstacles might preclude the realization of the ideal situation, but important is that the ideal warrants a consensual, rational outcome.) Where Habermas puts emphasis on the ideal deliberative procedure and on communicative rationality[2] in order to obtain consensus, Rawls justifies principles and institutions on the grounds that they would have been agreed unanimously in a hypothetical deliberative situation. Furthermore, Rawls (1993) introduces the idea of an *overlapping consensus* (on a shared conception of justice) in the public sphere, relegating disagreements to the private sphere – where a plurality of different and irreconcilable comprehensive views coexists. We will not get into the details or criticisms of the impressive oeuvres of Rawls or Habermas here. The main point I want to emphasize is that they both see politics as fundamentally seeking (and finding) a stable consensus to accommodate pluralism. What they have in common is that they believe in encountering a stable consensus surmounting the differences (between different reasonable comprehensive doctrines) – and consider it imperative.[3]

Furthermore, the consensus sought by deliberative theorists is interpreted as in the interest of all. As Seyla Benhabib articulates: "According to the deliberative model of democracy, it is a necessary condition for attaining legitimacy and rationality with regard to collective decision making processes in a polity, that the institutions of this polity are so arranged that what is considered *in the common interest of all* results from processes of collective deliberation conducted rationally and fairly among free and equal individuals" (1996, p. 69, *my italics*).

Having highlighted the ideal of consensus and the common interest as central components of the theory of deliberative democracy, we will now turn to a democratic theory critical of both.[4]

Agonistic pluralism and dissent on consensus

In this section, we introduce a theory of democracy, *agonistic pluralism*, as elaborated by Chantal Mouffe, which criticizes the theory of deliberative democracy sharply; it questions the ideal of consensus, emphasises that the dimension of power is ineradicable, puts contestation and antagonism central and values pluralism positively, cf. "the type of pluralism that I am advocating gives a positive status to differences and questions the objective of unanimity and homogeneity, which is always revealed as fictitious and based on acts of exclusion" (Mouffe 2000a, p. 19).

We would like to focus on two aspects of the criticism in particular. First, that the consensus cherished by the theory of deliberative democracy is *de facto* not inclusive but oppressive. Following the Habermasian ideal of deliberation, it is necessary that the collective decision-making processes are so organized that the results are equally *in the interest of all*, representing an impartial standpoint (cf. Benhabib 1996, p. 69). But agonistic pluralists claim that the consensus decision-making (in the common interest of all) conceals informal oppression under the guise of concern for all by disallowing dissent.[5] Or, as Iris Young puts it: "When discussion participants aim at unity, the appeal to a common good in which they are all supposed to leave behind their particular experience and interests, the perspectives of the privileged are likely to dominate the definition of that common good" (Young 1996, p. 126). Deliberative theorists postulate the availability of a consensus without exclusion (in the interest of all), an important point we will return to in relation to scientific pluralism (in the section entitled 'Scientific pluralism and dissent in science').

Second, the theory of deliberative democracy eliminates conflict and fails to keep contestation alive. Agonistic pluralism points out the theory of deliberative democracy can accommodate pluralism only by a strategy of depoliticization. This impedes us from acknowledging the nature of the political struggle – the *political* – according to Chantal Mouffe; it is *the end of politics*.[6] We have to acknowledge power, according to Mouffe, instead of ideally eliminating power and conflict: "In a democratic polity, conflicts and confrontations, far from being signs of imperfection, are the guarantee that democracy is alive and inhabited by pluralism. This is why we should be suspicious of the current tendency to celebrate the 'end of politics'. (...). This is why its survival depends on the possibility of forming collective political identities around clearly differentiated positions and the choice among real alternatives" (Mouffe 2000b, p. 4). Thus, political contestation should be a continuous practice, avoiding the oppressive consensus (seemingly in the interest of all) and acknowledging the plurality of values and interests (valued positively).

Thus, Chantal Mouffe's version of agonistic pluralism takes social life to be inherently antagonistic – without the possibility for any final reconciliation. She suggests that, rather than seeking an impossible consensus, a democracy should enable and institutionalise a plurality of *antagonistic* positions in order to create *agonistic* relations. Agonistic relations differ from antagonistic relations to the extent that antagonism denies every possibility of a (stable or shaky) consensus between the different positions; the 'other' is seen as an enemy to be destroyed. The idea is not to find a *common ground*, but to conquer more ground, to annex or colonize. Every party wants – in the end – to get rid of pluralism and install its own regime. Contrary to these antagonists, agonistic pluralists do cherish some form of common ground, a 'conflictual consensus', which we will discuss in the next section. In this sense, agonists 'domesticate' antagonism, so that opposing positions confront one another as *adversaries* who respect their right to differ, rather than *enemies* who seek to obliterate one another.

Furthermore, Mouffe has been pointing at the relation between *antagonism* and the *consensual deliberative theorists*; citizens that do not have the possibility of choosing among real alternatives within a democratic system (which is the case when it is too consensual, according to Mouffe), turn to right-wing populists in their dissatisfaction – populists that exploit this dissatisfaction and present themselves as the only alternative to the establishment. The antagonistic relation, as analysed by Mouffe, is one between outsiders and the establishment, the insiders, presented as one homogeneous group. Whether these inimical relations are mainly the effect of the insiders actually grouping together excluding outsiders, or the outsiders presenting the insiders as grouping together in order to get bigger themselves, is not always clear. According to Mouffe, it seems that the insiders are to 'blame'; she has argued that "the growth of right-wing populist parties in many European countries should be seen as the result of the type of 'politics at the centre' which has become predominant in recent years. With the blurring of the frontiers between left and right, conflicts cannot be expressed any more through the traditional democratic channels hitherto provided by party politics. Contrary to the claim made by Third Way theorists that the adversarial model of politics has become obsolete, this lack of a truly agonistic debate between adversaries does not represent progress for democracy (which would supposedly have become more mature), but something that can really endanger democratic institutions" (Mouffe 2005b, p. 228).

In order to further clarify the different models of democracy as presented by Mouffe, we illustrate how they result in different political party constellations. An *agonistic* democracy sees a confrontation among (clearly differentiated) adversaries, for instance, among a liberal-conservative, social-democratic, neoliberal and radical-democratic party. The *antagonistic* democracy sees a struggle between enemies who do not have any shared principles, for instance, between populist right-wing/anti-Establishment parties and the 'consensus at the centre', for example, the Third Way politics. The *consensual* democracy sees a consensus at the centre, for instance, the Third Way as consensus, attempting to transcend the left/right opposition, both old-style social democracy and neoliberalism. According to Mouffe, this consensual democracy will eventually result in the same constellation as the antagonistic democracy, given that the lack of real alternatives will engender dissatisfaction exploited by right-wing populism and anti-establishment parties.

In conclusion, agonistic pluralists aim at avoiding the antagonistic constellation, looking for ways to transform antagonism into agonism. The political system should provide a framework that deals with conflicts through an agonistic confrontation among adversaries rather than an antagonistic struggle between enemies (cf. Mouffe 2000a, p. 117). The agonistic confrontation will then keep democratic contestation alive (contra consensual deliberative theorists), without jeopardizing democracy (contra antagonism).

Towards a consensus on dissent in democratic theory?

Where deliberative democrats promote consensus-building processes, assuming it will serve the interests of all, agonistic pluralists accentuate that the appeal to the common good and consensus decision-making of deliberative theorists can often be oppressive, equating the common good with the interests of the more powerful, thus sidelining the concerns of the less powerful as well as eliminating the political and pluralism (the latter being the endorsement of the existing plurality of interests and values). This disagreement or dissent on the consensus ideal raises a couple of questions: Can we possibly find a way to reconcile the consensus ideal and pluralism? Can we find a form of consensus on dissent? Are there limits to pluralism and dissent? Does the plea against the consensus ideal inevitably result in *anything goes*?

At least three different roads have been explored to deal with the dissent on consensus. First, one could distinguish a nonoppressive from an oppressive consensus, as done by Steve Fuller (2007); he distinguishes a hegemonic and an emancipatory form of consensus, inclining towards the latter.[7] Could this be more than an unworldly ideal? Where does this emancipatory consensus, for instance, place the often conflicting interests of different emancipatory movements? Second, the consensus ideal and pluralism could be reconciled by accepting and valuing pluralism at the simple level of interests and values in coexistence with a consensus at the metalevel. This is suggested by Dryzek and Niemeyer (2006), but one has to conclude that their metaconsensus is very Habermasian – "produced by relatively uncoerced dialogue" (p. 647) with a strong belief in rationality (and, therefore, has to deal with the same problems already discussed in the section entitled 'Agonistic pluralism and dissent on consensus'). Third, dissent on consensus can result in a form of consensus on dissent like the *conflictual consensus* presented by Mouffe. The conflictual consensus differs from the former approach of the metaconsensus, because it is not static; the emphasis is put on the shakiness, temporality and precariousness of this common ground, endorsing contestability and pointing out that a stable consensus on pluralism eventually denies pluralism. It is within the conflictual consensus that the confrontation among clearly differentiated adversaries, not enemies, takes place: An adversary is "a legitimate enemy, an enemy with whom we have in common a shared adhesion to the ethico-political principles of democracy" (Mouffe 1999, p. 755). However, there is dissent about the interpretation, the meaning and implementation of these ethicopolitical values and principles, one that could not be resolved through deliberation and rational discussion. That is the antagonistic element in the conflictual consensus.

Returning to the questions raised in the beginning of this section, a way of reconciling the ideal of a (meta)consensus with pluralism can thus be the creation of the conflictual consensus stipulating the dynamic 'rules of the game'. This consensus does put limits to pluralism, limits that are not decided on moral or rational grounds, but on political grounds – no democratic society can exist without some form of exclusion, some way of putting limits to what is legitimate: "The pluralism that I advocate requires discriminating between demands which are to be accepted as part of the agonistic debate and those which are to be excluded. A democratic society cannot treat those who put its basic institutions into question as legitimate adversaries. The agonistic approach does not pretend to encompass all differences and to overcome all forms of exclusions. (...) To be sure, the very nature of those institutions is also part of the agonistic debate, but, for such a debate to take place, the existence of a shared symbolic space is necessary" (Mouffe 2005a, pp. 120–121). The lack of such a shared symbolic space (i.e., the conflictual consensus), of agonistic channels through which grievances can be expressed, tends to create the conditions for the emergence of extreme, antagonistic conflicts; conflicts that can most likely be avoided if the agonistic framework with legitimate political channels for dissent is available, according to Mouffe. Finally, even though there is a plurality of interpretations of the shared ethicopolitical principles of democracy, the agonistic framework does avoid that pluralism results in *anything goes*; for instance, the populist right-wing parties are considered not living up to the conflictual consensus, not sharing the common ethicopolitical principles of a pluralist democracy.

Scientific pluralism and dissent in science

Having discussed the differences between *models of democracy* concerning the ideal of consensus, dissent and pluralism, we will now elaborate the symmetries with *models of science* and their ways of dealing with the scientific consensus, dissent and pluralism. Especially – but not only – in the social sciences one often encounters competing theories, both intra- and interdisciplinarily. Upon closer scrutiny, several theoretical clusters can be distinguished across the social sciences, that is, because some theories share a similar ontology, a form of explanation, a concept of rationality, ideas about freedom, general assumptions about society and history, the necessity of critique and so on. It might result in distinguishing behaviourism, functionalism, rational choice theories, structuralism (and its more recent versions), culturalism, (socio-)biological theories, critical theories and so on. Inevitably, this situation raises questions such as the following: What is the source of this plurality? How to deal with this plurality? Should the plurality of theories, models and explanations in science ideally be replaced by a consensus? Does this presuppose the possibility of monism, a single, complete, and comprehensive account of the world (or of the phenomena investigated by science) based on a single set of fundamental principles (cf. Kellert et al. 2006, p. x) or are there ways to reconcile the consensus ideal with scientific pluralism?

Let us start addressing these questions with an inquiry into different interpretations of scientific pluralism – as the normative endorsement of plurality – and their relation to dissent. I will discuss three different interpretations[8]:

a. **Pluralism, no dissent:** there are many different systems of representation for scientific use in understanding nature, none of them complete, and jointly consistent
b. **Pluralism and dissent:** there are many different systems of representation for scientific use in understanding nature, none of them complete, and they may be irreconcilable or noncongruent
c. **Pluralism/plurality (Yes, No) and dissent (No, Yes):** there are many different systems of representation for scientific use in understanding nature, all of them (aim to be) complete, and (presumed to be) irreconcilable[9]

Philip Kitcher's version of pluralism is a good instance of the first interpretation of pluralism. Helen Longino's interpretation illustrates the second. Debating with Longino, Kitcher (2002) sums up four claims of his scientific pluralism: "(1) there are many different systems of representation for scientific use in understanding nature; (2) there is no coherent ideal of a complete account of nature; (3) the representations that conform to nature (the true statements, the accurate maps, the models that fit parts of the world in various respects to various degrees) are jointly consistent; (4) at any stage in the history of the sciences, it's likely that the representations *accepted* are *not* all consistent" (Kitcher 2002, pp. 570–71.) (Notice that I adopted parts of the wordings of Kitcher for describing the three versions of pluralism, supra.)

Longino characterises Kitcher's (2001) pluralism as *modest*. Kitcher does not presume monism; he allows different systems of representation, representing nature in a way that lives up to our interests, and he is cognizant of the incompleteness of these representations. But, according to Longino (2002b, p. 575): "No sense, however, can be made by Kitcher of my suggestion that equally successful representations may be irreconcilable or non-congruent in any non-redundant way. Hmm. I countenance pluralism because

I countenance the possibility of different equally defensible background assumptions facilitating inferences to quite different and irreconcilable, even non-mutually-consistent, representations of what is pre-theoretically identified as the same phenomenon." This could be the case, for instance, because: "Different equally defensible parsings of an ontological space can result in approach B treating the causally significant factors of approach A as undifferentiated portions of the environment of causally significant factors for B and vice versa" (Longino 2002b, p. 575).

Concerning Longino's interpretation of pluralism, it is important to mention that she does not exclude the possibility of monism (Longino 2002a, p. 75). She wants to avoid metaphysical prejudices and promote empirical study, that is, identifying in which situations multiple models or explanations are required and studying how plurality in the particular sciences ought to be understood (cf. Kellert, Longino and Waters 2006). She describes her position as "a commitment to avoid reliance on monist assumptions in interpretation or evaluation coupled with an openness to the ineliminability of multiplicity in some scientific contexts. (...) the plurality in contemporary science provides evidence that there are kinds of situations produced by the interaction of factors each of which may be representable in a model or theory, but not all of which are representable in the same model or theory (...) A more complete representation of some phenomena in contemporary science does require multiple accounts, which cannot be integrated without loss of content" (Ibid., pp. xiii–xiv).

This is very different from Kitcher's approach; he presumes the different systems of representations to be jointly consistent, there is a way to integrate the plurality of approaches or accounts in a scientific discipline even though at "any stage in the history of the sciences, it's likely that the representations *accepted* are *not* all consistent". Kitcher's modest pluralism subscribes to the plurality present in contemporary science, but considers it to be resolvable (at least) in principle (cf., Kellert et al. 2006, p. xi). Where does this conviction of consistency stem from and how to justify it; what are the benefits of denying the possibility of nonreconcilability, nonmonism? Here we want to point at the symmetry with the postulation of the availability of a consensus without exclusion by deliberative theories, mentioned above.

Another form of modest pluralism, of interpretation (a) of pluralism, is Sandra Mitchell's (2002, 2004); Mitchell acknowledges that the world is patchy but she sees one model or theory explaining phenomena in one patch and a different model or theory explaining phenomena in another patch, cf. "However complex, and however many contributing causes participated, there is only one causal history that, in fact, has generated a phenomenon to be explained" (Mitchell 2002, p. 66). So, there are comprehensive, complete accounts (of the causal history) for each particular phenomenon; (local) integration is presumed to be always possible and desirable: "The 'levels of analysis' framework describes the territory of pluralistic investigations, but it is only by integration of the multiple levels and multiple causes, including attention to the diverse contexts in which they occur, that satisfactory explanations can be generated" (Mitchell and Dietrich 2006, p. S78). Hence, could one not conclude that Kitcher's and Mitchell's approaches are sophisticated forms of monism?

We can now move Mouffe's grid of models of democracy (i.e., consensual, antagonistic and agonistic) inside the debate on scientific pluralism as presented here, articulating different models of science and scientific pluralism. Kitcher's and Mitchell's approaches might be characterized as consensual, ultimately presuming a consensus like consensual deliberative theorists. Hence, the first interpretation of pluralism and dissent can be labelled a form of *consensual pluralism*, or *consensual mainstreaming* (cf. infra). The second interpretation, as represented by Helen Longino's interpretation of scientific pluralism, can handle irreconcilability, neither presumes consensus nor the eliminability of multiplicity and accepts the possibility of 'different equally defensible' perspectives.[10] We can label this interpretation as *agonistic pluralism*, or *agonistic engagement*, because it emphasises (non-consensual) difference as well as highlights the importance of dissenting and adversarial engagement.[11] The third interpretation of scientific pluralism can be labelled *antagonistic exclusivism*; it does 'realise' the plurality of systems of representation, but denies every possibility of consensus, engagement or interaction between the (allegedly) inimical positions. One could understand plurality as a sum of exclusive *monisms*, and the relation between these *monisms* might be one of *dissidence* – i.e., a strong form of dissent – inimical opposition or one of indifference, isolation, *anything goes*. One could also think of the (metaphysically or otherwise) impossibility to interact (the idea of incommensurability).[12]

Consensus and dissent in the social sciences

Presenting some controversies in the social sciences, in particular in economics and sociology, we will analyse the presence of the idea of a scientific consensus and make the three versions of pluralism more concrete as well as illustrate the idea of agonistic pluralism in the social sciences. Talking about theories in the social sciences, consensual pluralists should aim for achieving a consensus theory under the conditions of plurality. This might be a comprehensive theory as a framework within which different theories peacefully coexist, or a synthesis, integration or unification of competing theories that gets the label of new consensus. I will first give a short example in sociology of the idea of the synthesis and then consider the controversy within economics between so-called orthodox and heterodox economists, which will give us the opportunity to further analyse the different versions of pluralism.

Sociology

Much of sociological theory in the 1980s and 1990s was centred on the so-called structure-agency (and/or micro-macro) problem, for instance, in the work of Margaret Archer, Jeffrey Alexander, Pierre Bourdieu, James Coleman and Jürgen Habermas. Anthony Giddens also tried to overcome the divide between structured-centred and agency-centred approaches, by developing a new unifying theory, that is, *structuration theory* (see Giddens 1984). It views society as a structuration process whereby human actions are simultaneously structured by society and structural determinants of society. As such, it resolves the tension between agency-centred, subjectivist approaches and structure-centred, objectivist approaches and transcends the dualism between micro- and macro-sociology building a new synthesis.

The same Anthony Giddens was engaged in a similar synthesising project in relation to politics. In his books *Beyond Left and Right* (1995) and *The Third Way* (1998), he attempts to transcend the left/right opposition, between the old-style social democracy and neoliberalism, after the demise of the communist model. Giddens's books provided the intellectual legitimacy for Tony Blair's New Labour Party and its Third Way. Chantal Mouffe refers to the Third Way of New Labour in Britain, led by Tony Blair, as an example of the 'end of politics'. The consequence of Blair's postpolitical, nonconflictual 'consensus at the centre' – and the reduction of politics to spinning – is a growing dissatisfaction with politics, a drastic fall in participation in elections as well as the success of the populist right-wing, according to Mouffe.

Returning to the 'intellectual guru' of the Third Way, Anthony Giddens, one can raise the question whether his new consensus theory in sociology is in the (epistemic) interest of all; whether all (epistemic) interests are just as well or better being served by the reconciling synthesis as they were by the replaced older theories – assuming that no epistemic benefits are lost in the emergence of a consensus theory. Upon closer scrutiny, we can demonstrate that the introduction of some new ontological furniture does not serve the interests of all; several epistemic interests are not addressed due to a lack of attention to different *forms of explanation,* losing interesting *explanatory information.*[13] Thus, Giddens's consensual approach turns out untenable as an epistemic standard, providing epistemically suboptimal results, hence *epistemically* undesirable.[14]

Economics

In this section, we will discuss the controversy within economics between orthodox (mainstream) and heterodox economists in order to analyse how different approaches within economics deal with plurality; how different versions of scientific pluralism are implicitly presupposed. It will analyse the desirability of some form of scientific pluralism maximizing epistemic fecundity – understood as maximizing the epistemic interests adequately addressed – and minimizing epistemically suboptimal strategies; the best option will turn out to be the agonistic engagement, or so will be argued in this section and the following one (see also Van Bouwel 2006).

Defining (the changing constitution of) orthodoxy and heterodoxy

As we will analyse the interaction between *orthodoxy* and *heterodoxy*, it seems appropriate to start with defining *orthodoxy* and *heterodoxy* – which will be done very concisely. First, the content of these labels follows the dynamics of the discipline, as John Davis (2008, pp. 353–54) notices: "neither 'orthodox' nor 'heterodox' inherently refers to any particular type of approach in economics; alternatively, any kind of approach in economics can be orthodox or heterodox depending on historical conditions, and indeed in the history of economics most major approaches have been both at one time or another, and not infrequently both at the same time, though in different locations." Second, we follow Davis (2008) in his categorization of the different approaches in contemporary of economics, that is

> *Orthodoxy* – neoclassical economics
> *New mainstream heterodoxy* – behavioural, experimental, evolutionary economics, neuroeconomics, evolutionary game theory, agent-based complexity economics
> *Traditional heterodoxy* – (old) institutional economics, Marxist economics (and radical economics), post-Keynesian economics, neo-Ricardian economics, social economics and socio-economics, Austrian economics, feminist economics, critical realism, and post-modernist economics

The development of the *new mainstream heterodoxy* might lead to a new orthodoxy in the near future, but for now this group of approaches is (unfortunately) often left out of the debate between the orthodoxy and (traditional) heterodoxy. Let us now zoom in on the features of that debate, taking into account that what follows is not a comprehensive review of the debate, but rather a review of some central contributions that helps us explicate the interpretations of scientific pluralism at play and the extent to which the interaction between different approaches in economics can be labelled agonistic.

The interaction between approaches

A first question is what form of interaction there exists between the different approaches in economics. Agonists and antagonists agree that boundaries and exclusion are unavoidable, but for the antagonist, these boundaries are reified and constitute a clear and permanent frontier between *us* and *them*, *friends* and *enemies*. Agonists do not consider different approaches as enemies, but as *adversaries*. Moreover, the boundaries with the adversaries are constantly challenged and changing; there is an engagement with *them*. How is the situation in economics?

The orthodoxy seems to present its account as complete, universal and 'value-free'. Hence, there seems to be no need for engagement with the traditional heterodoxy, and there seems to be no reaction to the heterodoxy's criticism. Concerning the heterodoxy, first, John Davis (2008, p. 360) notices the limited interaction between the new mainstream heterodoxy and the traditional heterodox approaches and recommends the latter to engage more actively with the former.[15] Second, another commentator on the orthodoxy-heterodoxy debate, David Colander, points at the lack of heterodox voices and ideas in the mainstream conversation and pleads for more "inside the mainstream heterodox economists" (cf. Colander et al. 2007). Third, the possibility of having agonistic engagement and interaction – keeping the agonistic channels open – between orthodoxy and heterodoxy is probably not helped by heterodox authors depicting orthodox economists as pathological cases (cf., Bigo 2008; at least a symmetrical argument for heterodox economists would be welcome).

There is more to economics than the orthodoxy. In that sense, an agonistic model can acknowledge the heterodoxy in economics as legitimate. On the other hand, what does agonistic pluralism imply for heterodox economists, how can they contribute to an interaction? If that is what they want, of course, they might defend antagonistic exclusivism (cf. infra).

The openness to dissent of the different approaches

A characteristic of agonistic pluralism as discussed above was the positive approach to dissent. Up to which extent are the heterodoxy and orthodoxy open to continuing dissent? Do they avoid (monistic/nonmonistic) metaphysical or explanatory prejudices that could exclude some contenders a priori and reify the boundaries between *us* and *them*?

One influential heterodox voice in the orthodoxy-heterodoxy debate is Tony Lawson's (cf. his 1997, 2003 and 2006). His views are grounded in an a priori ontological conviction referring to the transcendental argument for social structures elaborated by Roy Bhaskar (see Van Bouwel 2004a, for an analysis and critique of Lawson's ontological fallacy). Lawson's ontological convictions leave little space for dissent. His views restrict the possibility of interaction to those approaches that share Lawson's ontological conviction (as I will mention later, you might consider his view on pluralism as an instance of the first version aiming for integration of the different heterodox views, just as well as an instance of the third version in his strong opposition to orthodox views, with a reified boundary between orthodoxy and heterodoxy).

As noticed by many commentators (including ones that consider themselves to be orthodox), in order to engage in a discussion or critical interaction with the orthodoxy a formal, mathematical model is a *sine qua non*, cf., David Colander et al.: "modern mainstream economics is open to new approaches, as long as they are done with a careful understanding of the

strengths of the recent orthodox approach and with a modelling methodology acceptable to the mainstream" (Colander et al. 2004, p. 492). Milton
Friedman does comment on the mathematization of (orthodox) economics
as well: "Economics has become increasingly an arcane branch of mathem-
atics rather than dealing with real economic problems" (Friedman 1999,
p. 137). Where does that leave the openness to dissent, the channels for
agonistic engagement?

Attention for the adversary's diversity or caricaturising the enemy

How do the orthodoxy and heterodoxy deal with diverse perspectives? Do
they consider the opponent as an enemy to be destroyed or as an adversary
whose existence is legitimate and must be tolerated? One could try to answer
these questions by analysing up to which extent the other (and boundaries
with the other) are challenged and changed, and whether this challenge
is based on caricatures of the other or starts from a more subtle analysis of
the opponent. In relation to the heterodoxy, one could, for instance, raise
the following questions: Does the heterodoxy pay enough attention to the
diversity of approaches within the orthodoxy? Does the heterodoxy pay
enough attention to the diversity of approaches within the heterodoxy?

Attempting to answer the first question, I return to Tony Lawson's work.
Lawson's presentation of the orthodoxy (or the mainstream) is very homogeneous; the diversity of orthodox approaches is not valued by Lawson, on
the contrary, he tends to make a homogeneous caricature or straw man out
of the mainstream which invites critics to reject it *en bloc,* as we have argued
before (cf. Van Bouwel 2004a; see also Reiss 2004).[16]

The second question asks whether the diversity within the heterodoxy is
valued. Some commentators do recognize and value the diversity and plurality of the heterodoxy, see, for example, Davis (2008). Others try to identify
a common trait or 'the essence' of the heterodoxy, for example, Tony Lawson
(2006): "I am contending that the essence of the heterodox opposition is
ontological in nature" (p. 493); and, "Thus I am arguing that, collectively,
heterodox economists are primarily motivated, in their opposition to the
mainstream, by ontological (not epistemological) considerations" (p. 498).
Can the diversity of the heterodoxy be valued within Lawson's attempt of
ontological integration?

Similar questions can be raised concerning the orthodoxy. You do have
the advocates of unification of the mainstream, for example, Herbert Gintis
(2007), actually unifying all of the behavioural sciences. These unifica-
tory attempts do not take into account the fact that different approaches
may serve different epistemic or nonepistemic interests (see Weber and Van
Bouwel 2002). They drive on their 'scientific' status, which they contrast
with more traditional 'unscientific' approaches in the behavioural sciences.
John Davis perceives something similar in the relation between orthodoxy and heterodoxy: "Essentially, the orthodoxy – heterodoxy distinction

allows economics to claim economics is scientific by dismissing heterodoxy as unscientific" (2008, p. 352). The relation to the heterodoxy (and its diversity) is then instrumental in the way that the heterodoxy's *unscientificness* supports the orthodoxy's *scientificness;* the heterodoxy functions as a constitutive outsider. Is it valuing diversity?[17]

The (im)partiality of approaches in economics

The last issue we want to address concerning the interaction between orthodoxy and heterodoxy is to what extent plurality is cherished – the plurality of clearly different adversaries with their own legitimate interests and values. This would go hand in hand with the awareness of the partiality of approaches in the orthodoxy-heterodoxy debate. Actually, we can only observe that the debate proceeds very much in *all-or-nothing*-terms; benefits and limitations of particular models, explanations, theories or the limited range of concepts and methods are seldom made explicit.

In a sense, Tony Lawson does pay attention to partiality, be it only to the partiality of (traditional) heterodox approaches (forgetting the various approaches of – what he calls – the orthodoxy): "Rather, I suggest that the most, and perhaps only, tenable basis for drawing distinctions between the various heterodox projects is *according to substantive questions raised or problems or aspects of the socio-economic world thought sufficiently important or interesting or of concern as to warrant sustained and systematic examination.* That is, I suggest that the separate projects be characterised according to the features of socio-economic life upon which they find reason continually to focus their study" (Lawson 2006, p. 499, *his italics*). Lawson makes the division of labour between post-Keynesians, institutionalists, Austrians, feminist economists and others heterodox approaches, explicit, focusing on their different concerns, substantive orientations and emphases, "in a manner that does not compromise their coherence as fruitful traditions in economics" (Ibid., p. 502).

Recalling the orthodoxy's claim of *scientificness* referred to above, acknowledging partial concerns or partial epistemic interests and values seems to be unlikely; the value-ladenness is hidden, the 'value-free' completeness presumed. Awareness of its own partiality, and of the legitimate interests and values of adversaries in agonistic engagement, seems completely absent.

The different interpretations of scientific pluralism

Having gone quickly over several aspects of the heterodoxy-orthodoxy debate, one gets an idea of the different versions of scientific pluralism implicitly presupposed. The orthodoxy seems to understand itself (and its own diversity and plurality) as the integrating consensus theory; the mainstream as consensus, cf. consensual mainstreaming. As concerns the heterodoxy, on the one hand, you have some heterodox voices that seem to imitate the orthodoxy in their ideas of integration and advocate the consensual version of scientific pluralism, aiming for a heterodox integration. On the other hand, some of the self-declared pluralists of the heterodoxy seem to consider such a heterodox integration as the new dominant approach to substitute the orthodoxy (the orthodoxy might then just be reduced to play the role of the constitutive outsider, a role now played by the heterodoxy); in that sense, they support the antagonistic version of scientific pluralism.

Notwithstanding the options taken by the heterodoxy, the orthodoxy could as well be criticized within an agonistic version of pluralism without having to rely on the consensual or antagonistic version of scientific pluralism. The characteristics and benefits of agonistic pluralism will be clarified in the next section – for economics, it will enable the adjustment of both the mainstream and nonmainstream approaches, foster a positive approach towards dissent and make us take into account underprivileged perspectives. Simultaneously, an agonistic framework might result in emphasizing the partiality and limitations of the different approaches, making explicit the concerns and values of neuroeconomists, behavioural economists, institutionalists, neo-classicists, feminist economists and other.

Towards a consensus on dissent in the social sciences?

The examples of sociology and economics have shown how the different models of democracy presented by Mouffe can be used to articulate the interaction between different approaches in the social sciences and the different understandings of scientific pluralism – how these different approaches (should) interact and can(not) coexist. Through the analysis of these different versions, we are looking for the framework for scientific pluralism, which is epistemically the most desirable, maximizing epistemic fecundity, taking into account the plurality of interests and values. The examples have illustrated the epistemic suboptimality of the consensual approach as well as the reifying, unproductive character of the antagonistic approach. Let us now clarify why agonistic pluralism is the most preferable version of scientific pluralism. Comparing *agonistic engagement, consensual mainstreaming* and *antagonistic exclusivism,* three issues – in line with our analysis of the controversy in economics – will be articulated, that is, a positive approach towards dissent, the attention for diversity and the partiality of knowledge.

A positive approach towards dissent

Agonistic engagement has a positive and engaging approach towards (continuing) dissent; it does consider the possibility that the plurality of approaches cannot always be integrated, or that the different approaches are not always consistent. Moreover, the interaction of the plurality of approaches – the more developed and evaluated, the better – generates better knowledge. Consensual mainstreaming, on the other hand, does presuppose some sophisticated form of monism in which dissent can only be understood as a temporary state, to be solved, hence, dissent is negatively valued, cf. Sandra Mitchell: "Different levels of analysis might target different partial causes, but will there be no competition between levels? (...) Answers to questions posed at the different levels of analysis cannot be satisfactorily answered without consideration of the other levels" (Mitchell and Dietrich 2006, p. S77). "The 'levels of analysis' framework describes the territory of pluralistic investigations, but it is only by integration of the multiple levels and multiple causes, including attention to the diverse contexts in which they occur, that satisfactory explanations can be generated" (Mitchell and Dietrich 2006, p. S78). Hence, integration does seem the sine qua non for satisfactory explanations and models. This clearly differs from agonistic engagement in which neither (monistic/nonmonistic) metaphysical nor explanatory prejudices are cherished which could exclude – or oblige us to integrate – some adversaries or contenders a priori (losing them as challengers of the own approach). Finally, antagonistic exclusivism lacks engagement and positive interaction, so there is no positive dissent which might improve the own account, only indifference or a strong inimical opposition between contenders.

The attention for diversity

As concerns the relations between mainstream and nonmainstream approaches, agonistic engagement neither overestimates the explanatory importance of scientific approaches in the mainstream nor dismisses approaches outside the mainstream too easily, given that it sees clearly differentiated adversaries as the most desirable constellation. Antagonistic exclusivism, on the contrary, is either completely in favour or radically against the mainstream. Consensual mainstreaming risks to be too mainstream, that is, overestimating the importance of mainstream approaches and neglecting 'outsiders' or minimizing important differences with outsiders in order to succeed in integrating the plurality of approaches – an integration that is presumed to be always possible.

As should be obvious from what preceded, the agonistic approach does value differences. The interaction between different clearly differentiated approaches, as modelled in agonistic engagement, could be a generator for better knowledge (using biases as resources). The more and more different the biases the better, hence, interaction with as many clearly differentiated adversaries as possible is imperative. Consensual mainstreaming does minimize or overlook important differences among scientific approaches. As consensualists, differences are being treated as something to be eliminated (in an integrative account), not as a resource for better knowledge production. Furthermore, the plurality of diverse biases is not dealt with, cf. supra. Finally, antagonistic exclusivism is interested in showing that one's own account is complete or comprehensive and/or cannot gain or profit from an interaction with, let alone taking into consideration, different accounts.

The partiality of scientific knowledge

The agonistic engagement of clearly different adversaries in a confrontation should make them aware of the different (possible) legitimate interests and make the limitations of one's own particular models, explanations, theories and the limited range of concepts and methods visible. As such, it can emphasize the partiality of scientific knowledge and avoid exaggerated beliefs about the comprehensiveness of scientific knowledge.[18] Given that antagonistic exclusivism presumes the completeness of the own account and does not interact, partiality will remain invisible. Finally, consensual mainstreaming does perhaps recognize some form of partiality, the lack of an overall comprehensive unified theory, but it sees impartial, comprehensive theories or explanations about specific phenomena (cf. Mitchell's integrative explanations) or understands different true explanations or theories of the same phenomenon as translatable variants (cf. Kitcher's translatability/ congruency of maps).

Advocating agonism

Having compared different frameworks within which the plurality of approaches within the social sciences can be understood, we want to advocate a framework based on agonistic pluralism through which clearly differentiated alternatives can be confronted, sustaining the dynamics of science, dissent and heterodoxy. Implementing agonism in the (philosophy of the) social sciences will foster the confrontation and comparison through which the (relative benefits of) different explanatory, ontological and methodological choices of different approaches can be made more explicit, and within which the (epistemic and nonepistemic) interests they serve can be made visible, rather than continuing the debate about the best (consensus) choice for all possible situations, serving the interests of all or providing the ultimate method.

Conclusion

Starting from the scientific consensus as an ideal, we have explored its affinity with theories of deliberative democracy. Chantal Mouffe's critique on the consensus ideal of deliberative theories in democratic theory provided us with models of democracy from which the scientific consensus could be questioned as well as alternative models of science and scientific activity elucidated. Advancing Mouffe's agonistic pluralism as a model of social science, and in particular a specification of scientific pluralism, not only optimizes the epistemic fruitfulness of social science in relation to the plurality of interests, it also provides us with a democratic framework for dealing with dissent and the plurality of approaches instead of promoting the establishment of an epistemically suboptimal consensus.

Notes

1. Cf. Michael Friedman (2001, p. 54n) relates scientific rationality with the work of Habermas and Rawls, both labelled as defenders of (be it variations of) deliberative democracy. Stephanie Solomon's starting hypothesis (in this volume) is that deliberative democracy arguably "approximates the social practices of scientists", their reason-giving and collective argument, the best.
2. "This concept of *communicative rationality* carries connotations that ultimately race back to the central experience of the noncoercively uniting, consensus creating power of argumentative speech, in which different participants overcome their initially subjective points of view, and, thanks to the commonality of reasonable motivated convictions, assure themselves simultaneously of the unity of the objective world and the intersubjectivity of their context of life" (Habermas 1981/84, vol. 1, Chapter 1, section 1.A.).
3. Though aware of the 'fact of plurality', one has the impression that both Rawls and Habermas seem to be a bit uncomfortable with plurality and do not value plurality and pluralism positively.
4. Many criticisms of the theory of deliberative democracy have been formulated (inter alia, deliberation is just talk; neglects the value of intuition and gut-feeling; group decision is often based on conformity; *esprit de corps*, groupthink, see, e.g., Femia 1996; Solomon 2006), but we will discuss one critical theory in particular, fruitful for the understanding of scientific consensus and pluralism, or so will be argued.
5. Rawls and Habermas present their model of democracy as the one that would be chosen by every rational and moral individual in idealized conditions. In that sense there is no place for dissent or disagreement – the one that disagrees is irrational or immoral; the political has been eliminated (cf. Mouffe 2005a, pp. 121–22); "Conflicts of interests about economic and social issues – if they still arise – are resolved smoothly through discussions within the framework of public reason, by invoking the principles of justice that everybody endorses. If an unreasonable or irrational person happens to disagree with that state of affairs and intends to disrupt that nice consensus, she or he must be forced, through coercion, to submit to the principles of justice. Such a coercion, however, has nothing to do with oppression, since it is justified by the exercise of reason" (Mouffe 2000a, pp. 29–30).
6. "This struggle among adversaries, which I have referred to as 'agonistic', is what democratic politics is really about and one should never try to put an end to the agonistic confrontation. Now, as we have seen, this is precisely what Rawls is attempting when he declares that one needs to go beyond a constitutional consensus because in such a consensus disagreement still exists concerning the status and content of the basic political rights and liberties and that this creates insecurity and hostility in public life. In fact, what his approach aims at erasing is the very place of the 'adversary'. This is why, if it was ever realized, his ideal democratic society would be one where the agonistic struggle has come to an end" (Mouffe 2005b, p. 228).
7. Fuller relates this distinction to science. He sees science's aspiration to a universal consensus as a secularized version of the universalist aspiration of the great proselytizing religions (2007, p. 10). Hence, unworldly?
8. This list does not claim to be exhaustive. Other lists of forms of pluralism can be found in, for instance, Mitchell (2004), Solomon and Richardson (2005), Kellert, Longino and Waters (2006).

9. I am taking the options plurality (yes) and dissent (no) on the one hand, and pluralism (no) and dissent (yes) on the other hand, together. The first option can be understood as a form of *indifference* or as presuming the *impossibility* to compare models (everyone has her own model and is not really interested in other existing models or presupposes that it is impossible to compare models (incommensurability), hence, we have plurality, but no normative endorsement of it). Following the second option, you can have a plurality of models, but each models presents itself as a new monist (the other models being the 'enemies to be crunched'). Both options have a lack of fruitful interaction in common.

10. Cf. Mouffe (2005a, p. 52): "Adversaries do fight – even fiercely – but according to a shared set of rules, and their positions, despite being ultimately irreconcilable, are accepted as legitimate perspectives."

11. I do not want to claim that Helen Longino would subscribe completely to Chantal Mouffe's political philosophy. One could claim that Longino's ideas are mainly inspired by deliberative theory (see, for instance, her fourth CCE-norm concerning equality and inclusion), and therefore not coinciding with agonistic pluralism. However, Longino does share Mouffe's criticism on the presumed consensus and the exclusion of irreconcilability. Moreover, Longino's social epistemology is not as outcome-oriented as most deliberative theorists are, but rather focussing on procedure, a procedure which is always up for reinterpretation – cf. her third CCE-norm – which could show affinities with the *conflictual consensus*, a point I want to elaborate in future research. Van Bouwel (2008) explores how *agonistic engagement* – the second interpretation of scientific pluralism above – could be made philosophically more explicit, focussing on Helen Longino's *Critical Contextual Empiricism* (CCE) norms.

12. This category of antagonistic exclusivism might contain 'false' and/or 'strategic' pluralists (cf., Van Bouwel 2004a). One could claim that the category is not a genuine form of pluralism, but rather a collection of monisms; do consider, though, that the competing accounts can function as a constitutive outsider, which might imply that an account aiming for dominance does prefer to maintain the plurality rather than establishing monism (cf. the section entitled 'Economics').

13. An elaboration of this argument can be found in Van Bouwel (2004b) discussing Christopher Lloyd's *Structurism – Towards the Reunification of the Social Studies*, as well as in other articles in which the importance of a plurality of forms of explanations is defended in order to maximally serve different possible interests against advocates of one single best form of explanation (see, e.g., Van Bouwel and Weber 2008; Weber and Van Bouwel 2002). Another sociologist/philosopher wrestling to subsume a plurality of approaches within one comprehensive theory is Jürgen Habermas (cf. Bohman 1999). The phenomenon of syntheses that have to serve as a new consensus in the field frequently pops up in the social sciences; I have discussed other examples of synthesis (in International Relations Theory) and integration (driven by *economics imperialism*) in Van Bouwel (2009).

14. I emphasise the *epistemic* undesirability here; one could however motivate one's critique against a consensus theory pointing at the *political* injustice it installs, cf. Young's and Mouffe's criticism of a consensual approach discussed above, that is, oppressive – though seemingly in the interest of all.

15. Davis's recommendation has to be understood in combination with his conviction that a new orthodoxy is being made and that some input of the traditional heterodoxy would be desirable.

16. Concerning Lawson's *caricature* or *straw man*, I quote Julian Reiss (2004, p. 321): "What I want to show in particular is that Lawson's attack on the mainstream is to some extent misguided. He sets up a straw man called mainstream economics, which, according to Lawson, is caught in a positivist trap, where scientific explanations are no more than deductions from law-like statements, and the latter represent event-regularities. Since social event-regularities are rare, the mainstream project must fail as an explanatory endeavour. Lawson comes to the rescue with a scientific methodology tied especially to this world. What I want to argue here is that the mainstream has to some degree already absorbed the lessons from earlier criticisms of positivism." And: "do not criticize a straw man to be found in the textbook" (Colander et al. 2007). Furthermore, the lack of attention for differences also shows in the way Lawson deals with Davis's category of *new mainstream heterodoxy* (cf. supra). Lawson seems to consider it as part of the mainstream (or orthodoxy) without it seeming to have any impact on its homogeneity. Every time he 'attacks' the mainstream (or orthodoxy), though, he does attack *neoclassical economics*. The approaches of the *new mainstream heterodoxy* unfortunately do not get his attention (or are conveniently forgotten?). About the heterogeneity of the mainstream see as well Esther-Mirjam Sent (2006).

17. As mentioned before, an account aiming for dominance might prefer to maintain the plurality and diversity rather than establishing monism, because of the role others play as constitutive outsiders. However, this does not seem to be the kind of diversity that is epistemically fruitful.

18. We want to emphasise that partiality is understood as a positive characteristic here, showing the acknowledgment of dependency on different questions, interests, values and contexts. However, relative to these different questions, interests, values and contexts partial knowledge can be complete (i.e., serve the epistemic interests in the most complete and best possible way)!

References

Benhabib, Seyla (1996) 'Toward a Deliberative Model of Democratic Legitimacy?' In: *Democracy and Difference: Contesting the Boundaries of the Political*, ed. Seyla Benhabib. Princeton, NJ: Princeton University Press, pp. 67–94.

Bigo, Vinca (2008) 'Explaining Modern Economics (as a Microcosm of Society).' *Cambridge Journal of Economics* 32, 527–54.

Bohman, James (1999) 'Habermas, Marxism and Social Theory: The Case for Pluralism in Critical Social Science.' In: *Habermas: A Critical Reader*, ed. Peter Dews. Oxford: Blackwell Publishers, pp. 53–86.

Colander, David, Richard P. F. Holt and J. B. Rosser, Jr., eds (2004) *The Changing Face of Economics: Interviews with Cutting Edge Economists*. Ann Arbor: University of Michigan Press.

Colander, David, Richard P. F. Holt and J. B. Rosser, Jr. (2007) 'Live and Dead Issues in the Methodology of Economics.' *Journal of Post Keynesian Economics* 30(2), 303–12.

Davis, John (2006) 'The Nature of Heterodox Economics.' *Post-Autistic Economics Review* 40.

Davis, John (2008) 'The Turn in Recent Economics and Return of Orthodoxy.' *Cambridge Journal of Economics* 32, 349–66.

Dryzek, John S. and Simon Niemeyer (2006) 'Reconciling Pluralism and Consensus as Political Ideals.' *American Journal of Political Science* 50(3), 634–49.

Femia, Joseph (1996) 'Complexity and Deliberative Democracy.' *Inquiry* 39 (3–4), 359–97.

Friedman, Michael (2001) *Dynamics of Reason. The 1999 Kant Lectures at Stanford University.* Stanford, CA: CSLI Publications.

Friedman, Milton (1999). 'Conversation with Milton Friedman.' In: *Conversations with Leading Economists: Interpreting Modern Macroeconomics*, ed. Brian Snowdon and Howard Vane. Cheltenham: Edward Elgar, pp. 124–44.

Fuller, Steve (2007) *The Knowledge Book.* Montreal: McGill-Queen's University Press.

Giddens, Anthony (1984) *The Constitution of Society: Outline of the Theory of Structuration.* Cambridge: Polity.

Giddens, Anthony (1995) *Beyond Left and Right.* Cambridge: Polity.

Giddens, Anthony (1998) *The Third Way: The Renewal of Social Democracy.* Cambridge: Polity.

Gintis, Herbert (2007) 'A Framework for the Unification of the Behavioral Sciences.' *Behavioral and Brain Sciences* 30, 1–61.

Habermas, Jürgen (1981/1984) *The Theory of Communicative Action (Part I).* Cambridge: Polity.

Kellert, Stephen, Helen Longino and Kenneth Waters, eds (2006) *Scientific Pluralism. Minnesota Studies in the Philosophy of Science 19.* Minneapolis: University of Minnesota Press.

Kitcher, Philip (2001) *Science, Truth, and Democracy.* New York: Oxford University Press.

Kitcher, Philip (2002) 'The Third Way: Reflections on Helen Longino's The Fate of Knowledge.' *Philosophy of Science* 69, 549–59.

Lawson, Tony (1997) *Economics and Reality.* London: Routledge.

Lawson, Tony (2003) *Reorienting Economics.* London: Routledge.

Lawson, Tony (2006) 'The Nature of Heterodox Economics.' *Cambridge Journal of Economics* 30, 483–505.

Longino, Helen (2002a) *The Fate of Knowledge.* Princeton, NJ: Princeton University Press.

Longino, Helen (2002b) 'Reply to Philip Kitcher.' *Philosophy of Science* 69, 573–7.

Mitchell, Sandra (2002) 'Integrative Pluralism.' *Biology and Philosophy* 17(1), 55–70.

Mitchell, Sandra (2004) 'Why Integrative Pluralism?' *E:CO* 6(1–2), 81–91.

Mitchell, Sandra and Michael R. Dietrich (2006) 'Integration without Unification: An Argument for Pluralism in the Biological Sciences.' *The American Naturalist* 168, S73–S79.

Mouffe, Chantal (1999) 'Deliberative Democracy or Agonistic Pluralism?' *Social Research* 66, 745–58.

Mouffe, Chantal (2000a) *The Democratic Paradox.* London: Verso.

Mouffe, Chantal (2000b) 'Politics and Passions: the Stakes of Democracy' (paper). http://www.politeia-conferentie.be/viewpic.php?LAN=N&TABLE=DOCS&ID=viewpic.php?LAN=N&TABLE=DOCS&ID=124, last accessed September 2008.

Mouffe, Chantal (2005a) *On the Political.* London: Routledge.

Mouffe, Chantal (2005b) 'The Limits of John Rawls' Pluralism.' *Politics, Philosophy & Economics* 4(2), 221–31.

Rawls, John (1993) *Political Liberalism.* New York: Columbia University Press.

Reiss, Julian (2004) 'Critical Realism and the Mainstream.' *Journal of Economic Methodology* 11(3), 321–7.

Sent, Esther-Mirjam (2006) 'Pluralisms in Economics.' In: *Scientific Pluralism*, ed. Stephen Kellert, Helen Longino and Kenneth Waters. Minneapolis: University of Minnesota Press, pp. 80–101.

Solomon, Miriam (2006) ' "Groupthink" versus The Wisdom of Crowds: The Social Epistemology of Deliberation and Dissent.' *The Southern Journal of Philosophy* 44, Supplement, 28–42.

Solomon, Miriam and Alan Richardson (2005) 'A Critical Context for Longino's Critical Contextual Empiricism.' *Studies in History and Philosophy of Science* 36, 211–22.

Van Bouwel, Jeroen (2004a) 'Explanatory Pluralism in Economics: Against the Mainstream?' *Philosophical Explorations* 7(3), 299–315.

Van Bouwel, Jeroen (2004b) 'Questioning Structurism as a New Standard for Social Scientific Explanations.' *Graduate Journal of Social Science* 1(2), 204–26.

Van Bouwel, Jeroen (2006) 'Elucidating Scientific Pluralism in the Light of Economics and Economists.' Paper presented at the *Erasmus Institute for Philosophy and Economics*, Rotterdam, December 2006.

Van Bouwel, Jeroen (2008) 'Ways of Dealing with Scientific Plurality: Inquiring Helen Longino's Epistemic Democracy.' Paper presented at the *6th International Network for Economic Method* conference, Madrid, 13 September 2008.

Van Bouwel, Jeroen (2009) 'The Plurality of Epistemic Interests in Scientific Understanding Exemplified in Political Science.' In: *Scientific Understanding: Philosophical Perspectives*, ed. Henk de Regt, Sabina Leonelli and Kai Eigner. Pittsburgh, PA: University of Pittsburgh Press.

Van Bouwel, Jeroen and Erik Weber (2008) 'A Pragmatic Defence of Non-Relativistic Explanatory Pluralism in History and Social Science.' *History and Theory* 47(2), 168–82.

Weber, Erik and Jeroen Van Bouwel (2002) 'Can We Dispense with the Sructural Explanation of Social Facts?' *Economics and Philosophy* 18, 25975.

Young, Iris Marion (1996) 'Communication and the Other: Beyond Deliberative Democracy.' In: *Democracy and Difference: Contesting the Boundaries of the Political*, ed Seyla Benhabib. Princeton, NJ: Princeton University Press, pp. 120–35.

7
Joint Commitment, Coercion and Freedom in Science: *Conceptual Analysis and Case Studies*

Alban Bouvier

This chapter deals with the ethics of group life in the sciences, if not directly with the policy of science that might evolve from it, and more precisely with the issue of democracy within scientific life.[1]

I uphold a "naturalized" conception of ethics in the moderate sense that I consider that a relevant formulation of moral norms (like the duty of respect towards others) has to be illuminated by a close analysis of the effective social life (although not reduced to it). This viewpoint is compatible with a "naturalized" conception of social epistemology, understood as a formulation of the norms of knowledge (like logical and empirical validity norms) close to the effective processes of scientific life (Goldman 1999, 2002; Thagard 1998a, 1998b).[2] Consequently, this ethical and epistemological study would like to contribute to the sociology of knowledge in Robert Merton's (1942) style as well, aiming at characterising the "ethos of science", that is not only the effective standards of scientific life but also the ideal moral norms of science and reciprocally.

My starting point is the idea that *contractualist models* are relevant both for the analysis of the effective structure of scientific groups, that is at the descriptive level, and for providing reasonable normative guides of scientific ethos, but to an extent that has to be carefully investigated. My main point is to address the general issue of the nature and the degree of freedom that is both *ethically desirable* and *pragmatically accessible* within a research team.

In the first part of this chapter, I set forth two very different kinds of situations regarding the freedom issue in science, although both can be considered as based on "joint commitments". In the second part, I argue that many situations that seem to be based on joint commitments, which imply reciprocal and interdependent commitments, are actually based instead on mere unilateral commitments and that those may involve different kinds of ethically illegitimate constraints (or coercion). In the third part, I compare the concepts of joint commitment (Margaret Gilbert) and positive liberty (Isaiah Berlin), both inherited from Rousseau's contractualism. Rejecting positive liberty as well as negative liberty as unrealistic, I argue for the significance of a third conception of liberty as "absence of domination" (Philip Pettit). I uphold that it is a reasonable ideal in scientific life.

Unconstrained and constrained joint commitments in scientific life

I first contend that recent contractualist models of groups like Margaret Gilbert's or Philip Pettit's may be useful for analysing effective scientific groups, therefore at the pure descriptive level, whatever normative may be the main concern of each of them. Nevertheless, I will mainly consider Gilbert's account in this first part of my analysis, because Gilbert has devoted an illuminating paper specifically to scientific life.[3]

Margaret Gilbert has implicitly put forward an alternative to Kuhn's account of paradigms.[4] For the most part of his account of scientific progress, Kuhn examined the transmission of scientific principles during periods of "normal science", when the scientists learn the principles by doing typical exercises and repeating classical experiments within the paradigm at least as much as by explicit teaching. In these situations, there might be no real space for discussion since discussion requires clear awareness of the principles at stake. Kuhn's account of scientific life fits in with witnesses' accounts like Heisenberg's. Thus, according to a recent historian of quantum mechanics, Mara Beller (1999), Heisenberg conceded in his preface to "his 1930 book, which he dedicated to the 'diffusion of the Copenhagen spirit', that 'a physicist more often has a kind of faith in the correctness of the new principles than a clear understanding of them'" (p. 39). Actually, to use J. Cohen's (1992) distinction between belief and acceptance as, respectively, a passive and an active mental process, one can argue that within Kuhnian "normal science", scientists often have "beliefs", whereas in revolutionary periods, when a new paradigm arises, scientists have to choose between two paradigms and "accept" one of them after having weighed pros and cons. Mara Beller's commentaries on Heisenberg's confidence makes this distinction particularly clear, even if Beller does not herself use Cohen's concepts: "Young physicists, who streamed into these centres [Copenhagen, Göttingen, Leipzig, Hamburg] from all over the world, were exposed automatically [= "belief"] to the new philosophy. Because they were more interested in calculating and obtaining definite results than in philosophizing, most of them simply adopted [= "belief"] the official interpretation without deep deliberation [= without "acceptance"]" (p. 39) (commentaries between brackets are mine). Such deliberation would have been necessary, on the contrary, if students had been exposed both, on the one hand, to either Bohr (who directed the Copenhagen Institute), Born (who was teaching in Göttingen), Heisenberg (who conducted research in Göttingen, then in Copenhagen, then in Leipzig) or Pauli (who conducted research in Göttingen, then in Copenhagen, then in Hamburg), and on the other hand, to Schrödinger, who developed an alternative conception of quantum mechanics in a different place (Berlin).[5]

Gilbert (2000) suggests that new paradigms usually emerge not from the only addition of multiple individual acceptances but instead from the "joint acceptances" and even "joint commitments" of scientists.[6] As is now well known, Gilbert's main goal has been to try to make sense of Durkheim's intuitions that there are collective ideas which can be different from individual ideas and that these collective ideas can exert a constraint upon the individuals in the sense that the individuals might not be free to *not* profess them. Gilbert (1989, 2006) suggests that, at least in certain cases, it might be comparable to a contractualist situation where certain people vote against a law but must nevertheless accept it as *their* law and obey it if the majority has voted in favour of the law. Everybody is "committed" to obey the law and, more exactly, is "jointly committed" because each of the voter contracts with each other by voting and votes only under the condition that everyone does the same (thus, their commitments are interdependent). Gilbert claims that there are many situations in which there is not a formal and explicit contract but nevertheless a kind of *implicit* contract. For example, somebody may speak on behalf of the group as a self-proclaimed spokesman in such a way that everybody feels jointly committed to this leader and the other members of the group just because they did not explicitly disagree when he spoke, even if they do not agree deeply (Gilbert, 1994).

Gilbert did not write much on scientific collective beliefs – or acceptances – and, in her 2000 paper, she was essentially interested in the specific issue of the positive role of outsiders on the growth of knowledge, insofar as group commitments around a paradigm can hinder scientific progress. Thus, she was not interested in the general relevance of the joint commitment model for investigating the social scientific life. I claim that Gilbert's extension of the contractualist model to situations where there is not an actual contract is especially interesting because it is much more realistic (Bouvier, 2008).[7] In fact, it is rare there be an explicit contract relating to a paradigm or a research programme in the sciences. It might be the case when there is a public manifesto and the content of the manifesto is voted on by all the members of the group that supports it or by a similar formal procedure. But manifestoes are rare in the sciences, in particular in the natural sciences, and when there are such things, they are rarely the result of a formal procedure.[8]

As Paul Thagard has noted, "In most scientific fields (…), there is no central social mechanism that produces a consensus" (Thagard, 1998b). Nevertheless, Thagard adds: "In medical research, the need for a consensus is much more acute, since hypotheses (…) have direct consequences for the treatment of patients." Then, Thagard refers to the consensus conferences that exist in many countries on health issues. Thus, in the United States, the National Institutes of Health regularly convene such conferences. These conferences are constituted by a panel of experts on a specific issue (e.g., given the new results in sciences, what is the most recommendable treatment of gastric ulcer?). The panels take a decision after a two- or three-day deliberation. Deliberating is justified in these medical contexts because ambiguous data are frequent. Thus, as Thagard has reported, many gastric ulcers were sensitive to antibiotics but not all and, furthermore, the antibiotic treatment needed a tri-therapy, which was much more costly than the previous therapy. Consequently, a consensus conference appeared to be quite useful in order to take the right decision on the best remedy for gastric ulcers. [9]

The situation is comparable in other domains such as economics, because practical advice is at stake as well. Thus, institutes (private institutes), such as the von Mises and von Hayek Institutes or the Mont Pélerin Society (all of them founded in line with the viewpoints of the Austrian School of Economics, to which I will later on return regarding other aspects), give political recommendations based on a certain conception of how economics functions, an issue that remains uncertain. In such cases, on which I don't want to elaborate in detail, one often encounters situations in which the members of a minority have to "jointly accept" (with the majority) the claims of the majority, although they personally disagree with these claims. Consequently, the minority members will necessarily feel the group as constraining. But this constraint is implied by democratic procedures and consequently might be encountered in any acceptance of a democratic procedure.

As Thagard states, in most fields the practical implications are not so evident, so that the situation is quite different. However, in this field as in others fields, one can encounter *implicit* contract situations. There may be sorts of informal deliberation between the members of a group, for example, on the general principles of a research programme. But, in these latter cases, even if one considers that, in principle, contracts require from each participant a clear awareness both of what is contracted and of the existence itself of a contract,[10] in fact situations are often so ambiguous that it may turn out that certain individuals happen to be committed against their deepest will, then more or less forced to commit.[11] I will take two historical examples here to set forth the plausibility of these situations in sciences and to more clearly distinguish between the two kinds of constraints or coercions (one relative to any joint commitment procedure and one not). Certain differences may seem psychologically or sociologically very minor, but I claim that they make sense ethically.

First, when Emile Durkheim and Marcel Mauss (1903) founded a new sociology of knowledge research programme in a memoir they signed together (the sociology of the fundamental categories of thought programme), although they did not really set forth their aims in a manifesto, their joint memoir nevertheless sounded like a manifesto. Besides, no politically oriented academic society or institute was founded on its theoretical ideas (unlike what happened in the Austrian School case), but nevertheless they appear to have been explicitly committed, as in a contract, with regards to this programme. Thus, they constituted a "group" or, as Gilbert says, a "plural subject". Furthermore, they very probably also discussed at length the content of this programme, although one does not have any direct evidence (e.g., *via* letters) of that. They very probably deliberated on the relevance of referring to such or such theoretical principle and promoting such or such scientific goal, in the sense that deliberation is an exchange of arguments leading to a decision. Consequently, the programme was coherent, which means in particular that it did not look like a mere juxtaposition of two more or less different programmes, as happens in a syncretic synthesis obtained as a result of a mere negotiation.[12]

In this case, if either Durkheim or Mauss might have felt constrained by the result of the common deliberation (to some extent possibly different from their deep thoughts), and even if Mauss, as a junior, was more susceptible to have renounced personal ideas than Durkheim as a senior,[13] at least one does not have any evidence that Mauss was *compelled* to "jointly commit" to any common programme with Durkheim.

Finally, this programme was meant to give orientations to a team of colleagues and younger researchers like Hubert, Hertz, Czarnowki and so on. These individuals did not participate in the writing of the programme, and neither explicitly added their name to the "manifesto". But, they joined the first two, mainly through participating regularly in the writing and editing of *L'Année Sociologique*. Thus, with them they were also (almost explicitly) jointly committed to this programme.[14] But there is no evidence either that they had been more or less forced to commit either to this specific programme or to any collective programme.

Thus, the French School is a particularly clear example of a group of scientists who intellectually seem to have been jointly committed, in the contractualist sense of this expression, either in an explicit (Durkheim and Mauss signing a joint memoir) or in a quasi-explicit way (the other members participating in the journal of the French School, *l'Année sociologique*). A particularly striking feature of this group behaviour is that they happened to say "we" when they expressed ideas of the joint programme, even in personal papers. In these contexts, they publicly meant that they felt committed with the other members of the group even while expressing their own ideas and that they also felt entitled to involve the other members in their writings, to some extent at least, given that these members were supposed to be reciprocally and interdependently committed to the same common research programme. Marcel Mauss, in a lecture given as the 1938 *Huxley Memorial Lecture*, wrote: "Vous verrez un échantillon – peut-être inférieur à ce que vous attendez – des travaux de l'école française de sociologie".[15] And he added: "Nous nous sommes attachés tout spécialement à l'histoire sociale des catégories de l'esprit humain" (Mauss, 1950, p. 334).[16] In this context, "we" is not a conventional way to say "I" but clearly refers to the whole "French school of sociology", as the context shows. In the same paper, Mauss summarized the contributions of Hubert, Czarnowki, Durkheim and Lévy-Bruhl to the joint enterprise.

I borrow the second example again from the history of the Copenhagen group in quantum mechanics, but I now focus on the famous 1927 lecture Niels Bohr gave in Como, Italy, at the International Volta Congress in which he put forward the "complementary principle". Speaking about the same period we have referred to above, Heisenberg wrote that there was a kind of specific "spirit" in Copenhagen in these late twenties. Beller adds that, in the early thirties, it was the case not only in Copenhagen but in Göttingen as well, and still slightly later on in Leipzig and Hamburg, where, respectively, Heisenberg and Pauli had been given professorships. At that time, this "spirit" appears to have been just a "passive" belief. But, it was not yet the case in September 1927, in Como (at the Volta Congress) or in Brussels in October of the same year (at the Solvay Congress), that is, at the very beginning of the so-called "Copenhagen school".

According to Mara Beller, at Como, Bohr did not only want to express his ideas but also the ideas of the Copenhagen group. This group was constituted of physicists working together in Copenhagen in the same institute or regularly meeting there, mainly Heisenberg, Pauli and Dirac. Bohr's talk also involved other physicists closely connected to the others, such as Max Born (Heisenberg and Pauli had been his assistants in Göttingen) or Jordan (another of Max Born's assistants). Bohr even wanted to reconcile the stance of his team with Schrödinger's, someone who was to some extent an intellectual enemy. Thus, it resembled a collective manifesto, with somebody speaking on behalf of the group. However, Bohr was not formally delegated by the group to take on this role, and the group members did not write the text together (as Durkheim and Mauss did) or deliberate on the different issues at stake (e.g., on the meaning of "complementarity" or on Niels Bohr's attempt to reconcile quantum mechanics and classical physics). But, as neither Heisenberg nor Pauli explicitly and publicly disagreed with Niels Bohr's lecture either at Como or in further papers, the complementary principle was perceived by the entire community as the Copenhagen School interpretation of quantum mechanics.[17] However, one knows explicitly from the correspondence of these authors and implicitly from their articles that actually some of them, especially Heisenberg, *did not* accept the complementary principle at all (Beller 1999; Howard 2004). Moreover, none of the junior scientists explicitly jointly committed to Bohr's programme by participating in a specific review (like the members of Durkheim's group) or a handbook. Thus, in this situation it appears that, unlike the young members of the French School of Sociology vis-à-vis Durkheim, Heisenberg and Pauli were *forced* – to some extent – to jointly commit to Bohr and to assume a collective paradigm (of which they did not personally accept certain ideas).[18]

Nevertheless, from Gilbert's point of view as we will soon see in more detail, this makes no real significant difference with the French School example because the public silence of the Copenhagen group members was to be interpreted as a joint agreement by every participant in the Congress so that the Copenhagen group members were jointly committed *exactly in the same sense* as the Paris group members. According to me, on the contrary, the point here is that, if the intellectual joint commitment itself (whatever its specific content may be) was obtained only thanks to a kind of coercion, this commitment raises specific problems.

Actually, one could argue that the involvement of the other Copenhagen and Göttingen physicists was not yet as clear as it was in the Paris group case. Thus, in his Como talk and in the different written versions that were published from this, Niels Bohr never used "we" (unlike Mauss, for example), as he could have done if he wanted to very clearly implicate the Copenhagen-Göttingen group as such.[19] However, one month later, at the end of the Solvay Congress, Heisenberg and Max Born gave a joint lecture in which they pronounced this famous statement: "We regard quantum mechanics as a complete theory for which the fundamental physical and mathematical hypotheses are no longer susceptible of modification" (Heisenberg and Born 1927, p. 437).[20] In the context, it was clear that "we" did not refer only to Heisenberg and Born but instead to the whole Copenhagen-Göttingen group. And because Bohr did not explicitly disagree with the content of this paper, the usual feeling was again that all the Copenhagen-Göttingen physicists shared the same conception of quantum mechanics.[21] At Solvay, it was Heisenberg and Born who were the auto-proclaimed spokesmen and who more or less *forced* the others members of the Copenhagen-Göttingen group (ironically including Bohr himself) to accept their own common view. Then, all the scientists were publicly jointly committed to a common conception as in Como, although not exactly to the same conception as in Como, because they did not explicitly disagree.[22]

What can be considered as a consequence of the coerced joint commitment in Como and of the related absence of deliberation between the group members is that the Como talk is notoriously obscure.[23] Admittedly there are several interpretations of this obscurity. The most classical one is that Bohr's ideas were not yet clear enough when he wrote this paper and that he gave the audience his still emerging thought both on the relevant interpretation of quantum mechanics and on the unification of contemporary theoretical physics.[24] But Mara Beller's recent interpretation is that Bohr tried to amalgamate his ideas with Heisenberg's, Dirac's and Pauli's. Because their ideas were incompatible on certain fundamental points (which was not the case, e.g., in the Durkheim-Mauss memoir on the sociology of categories), Niels Bohr's amalgamating could not be but an incoherent syncretic synthesis.

To conclude the first part of this investigation, I will just state again that there is a pretty clear opposition between a freely agreed joint commitment (within the Paris School of Sociology) and a forced joint commitment (within the Copenhagen School of Quantum Mechanics). Although from Gilbert's viewpoint this does not make a relevant difference, one can argue that many scientists would not find Niels Bohr's attitude at Como (as described above, following Beller's interpretation) or Werner Heisenberg's and Max Born's at Solvay quite fair in comparison with Emile Durkheim's relation to Marcel Mauss and the members of the Paris group. I share this feeling and, in the second part of this chapter, I would like to focus on the ethical dimension of Gilbert's conception of joint commitment a little more directly.

Joint commitment, unilateral commitment and illegitimate constraints

To my knowledge, Gilbert has not elaborated at length on the previous normative problem (either in scientific life or elsewhere), and when she has addressed it, it was in relation to a closely related although different problem. It deals with Hobbes's and Rousseau's conceptions of social contract. In fact, provoking Rousseau's indignation, Hobbes's viewpoint left the door open to an ethical justification of domination and slavery – as far as this relation is "jointly accepted" (and that each contractant, either master or slave, feels "jointly committed").[25]

Commenting on Hobbes's conception of social contract, David Schmidtz (1990), whom Margaret Gilbert (1993) quoted, wrote: "I think it is more charitable to Hobbes to read his discussion as a purely descriptive account of the possible ways in which sovereigns can actually emerge, with no normative implications intended" (Gilbert 1993, p. 310, n. 44). But, responding to Schmidtz, Gilbert added after Schmidtz's quotation: "My dispute is with 'no normative implications'. No moral implications of a certain sort, perhaps. But (...) any genuine agreement has its own normative weight – as the conquering sovereign may well discern" (p. 310, n. 44).[26] I agree with Gilbert on the fact that *there is still a normative constraint* or, more exactly, that *people may feel committed* (or may *feel* the "normative weight" of the agreement), even when there is constraint or coercion on the commitment itself, that is, when participants have been to some extent compelled to jointly commit. And I find her viewpoint particularly astute and right sociologically speaking.[27] But, regarding the moral issue, I find Gilbert much too ambiguous. In fact, she seems only to concede that the commitment might not have moral legitimacy ("perhaps") and even to suggest that the same commitment might nevertheless have certain moral legitimacy, although of another sort than those Schmidtz seems ready accept ("no moral implications of a certain sort"). According to me, Gilbert's response remains much closer to Hobbes's ambiguous viewpoint than to Rousseau's intuitions, which I consider (probably like many other people) as the genuine moral intuitions. I will try to make this point more explicit regarding the scientific life case. Again, historical examples can help us to understand what is really at stake in these contexts.

Actually, to be perfectly clear, one has to distinguish carefully between two kinds of constraining situations. What might be revolting is a situation where the participants in a joint commitment are more or less forced to jointly commit, as at Como, that is a specific contractualist *procedure*. This was the matter in the first part of this chapter. But it might be also the *content* of the contract, especially when this contract implies entirely imposed constraints (slavery is an extreme case), something Philip Pettit calls "domination". This is the matter of Hobbes-Rousseau controversy. Not surprisingly, the two constraints might find themselves entangled in historical cases.

But a careful investigation suggests that joint commitment, therefore reciprocal and interdependent commitments, may be less frequent in sciences than Margaret Gilbert seems to believe (and that Gilbert's concept of joint commitment may lead one to think), whereas *unilateral commitments* may be much more frequent. Scientific groups are generally constituted of junior and senior researchers, so that it is likely that the commitment of each member with regard to a set of principles will *not* be reciprocal and interdependent: the senior will more likely expect a commitment from the junior to a set of principles without feeling committed with him to this set of principles.[28] As some examples will illustrate, the expectation of such commitment might be very constraining. Consequently, the examination of ethically illegitimate commitments deserves to be extended to these unilateral commitments. One will observe constraints both on the commitment content and on the commitment procedure (which this time means constraint to unilaterally commit). I will in succession investigate two different examples illustrating the two different situations. To facilitate comparisons, I chose these examples in scientific communities that I have already considered.

Let us first refer to the relations Bohr had years later after the Como conference with his students or young researchers. Beller insists on the fact that Bohr was now considered as a "hero" and was surrounded by a true cult to the extent that the young researchers did not feel able to think differently from Bohr and tried to interpret his most obscure papers as if they expressed the hidden truth of the world. Thus, according to Mara Beller, Weizsäcker once noted, after having met Bohr and understood hardly anything, "What must I understand to be able to tell what he meant and why was he right? I tortured myself on endless solitary walks"

(Beller 1999, p. 275). Weizsäcker's reaction seems to be pretty irrational because Weizsäcker seems *a priori* to eliminate the possibility that Bohr could have been wrong.[29] Beller comments: "The question was not: Was Bohr right? or To what an extent was Bohr right? or On what issues was Bohr right? But quite incredibly, What must one assume and in what way must one argue in order to render Bohr right?" (Beller 1999, p. 275). Thus, it looks like an *entire* submission to one another's thought and therefore a complete loss of freedom of thought, whether this other scientist was really aware or not of imposing his thought.[30]

Mara Beller adds that certain scientists (possibly including Weizsäcker himself) felt guilty for not thinking like Bohr. "Bohr's unpublished correspondence discloses the overwhelming guilt experienced by those physicists who dared to challenge him" (Beller 1999, p. 274). Experiencing guilt means that sorts of commitments were violated. But the first issue is to know if these were joint commitments between Bohr and his students (or junior scientists) or instead only students' (or junior scientists') unilateral commitments to Bohr's principles. In the former case, it would have meant that these students would have felt committed to the principles only to the extent that Bohr reciprocally seemed committed to the same principles, which is plausible. But the idea of joint commitment also implies that Bohr would have felt himself committed to these principles under the conditions that his students seemed committed as well (interdependence), which is quite unlikely. Consequently, although there is no absolute evidence of any kind, unilateral commitments only of the students are much more likely than (reciprocal and interdependent) joint commitments of everyone. And these cases are situations in which Bohr seems to have more or less consciously imposed his views.

Bohr's case is surely an extreme case (characteristic of charismatic leaders), but extreme cases of this kind are not rare in the social sciences. I would like to refer to another case chosen in a group I have also already referred to, not only because this example is less extreme but also because it illustrates another possible aspect of unilateral commitments hidden under apparent joint commitments. It is the case of Ludwig von Mises, one of the main representatives of the Austrian School of Economics, especially when he emigrated to the United States during the Second World War (and long before the foundation of the von Mises institutes by his followers). Mises was allegedly very authoritative with everybody (Caldwell 2004). But for years he entirely rejected one of his closest and most brilliant students, Fritz Machlup, just because the latter publicly defended scientific views (more exactly epistemological views) different from his. Mises is reported to have been indignant on this occasion.[31] This indignation reveals that he thought Machlup was committed to the same theories as he himself was.

Actually, one could argue that if there was a joint commitment to certain theoretical principles, for example, the Austrian principles (methodological individualism, subjectivism and focus on processes more than on equilibrium states, Boettke 1994b) or to certain more specific principles characteristic of the Mises microcommunity (see below), Mises might have been right to reproach Machlup. But the first point is to know whether Mises really felt jointly committed to these principles with Machlup and possibly with other junior scientists. Because there is no evidence of such a joint commitment feeling (unlike what emerges from the French School of Sociology materials, for example), it is much more psychologically plausible to think that Mises did not feel himself jointly committed to these principles, although he expected that Machlup felt committed to them. On the other hand, according to Fritz Machlup's memories (Machlup 1981), Machlup seems to have understood *he should have felt committed* to these principles only when Mises heaped reproaches on him.[32] Thus, the commitments at stake were at least not "common knowledge" with respect to Mises and Machlup, which would have required that the commitments be "out in the open" with respect to them (Gilbert 1996, p. 198).

But, what is specifically interesting in this case is that the "clash" between Mises and Machlup happened on the significance of one specific idea: Mises's idea that the rationality principle, taken as the most relevant principle in social sciences by many scientists, was analytical, that is, that human action has to be considered as *a priori* rational. For Mises, if certain behaviour was not rational, it was not specifically human and did not require specific explanation from the social sciences, but if it was human, it was rational by definition (Mises 1966). But as Rizzo (1990) has brilliantly shown, what can be discussed – even if one accepts the rationality principle as the "core" of a research programme – is the content of the notion of rationality. And the various meanings this notion can take constitute the "protection belt" of the programme, to use Lakatos's terms. The clash between Mises and Machlup took place when Machlup suggested such a modification of the rationality principle's meaning, in the light of the growing empiricist criticisms against Mises's "a priorism" (Boettke 1994a). It is surely a case where collective deliberations leading to decisions as to the relevance of the modification of certain principles of a paradigm would have made sense. But it was not what happened under Mises's "reign," because Mises seems not to have been able to accept any modification of the supposedly shared principles.

In this part, I have set forth two kinds of what I consider as illegitimate constraints imposed upon members of a scientific community within the context of commitments that seem at first sight to be joint commitments but reveal themselves to be more plausibly mere unilateral commitments of junior scientists. In one case, the constraint is rather on the content of the commitment: Weizsäcker seems to have almost abandoned confidence in his own thought, entirely submitting it to Bohr's. In the other, the constraint is rather on the procedure: Machlup was reproached by von Mises for either having violated a (unilateral) commitment or not having felt committed since Machlup wanted to initiate modification of the principles.

Now, at this point, I find it useful to introduce more astute moral or ethical conceptions of freedom and liberty than those I have until now implicitly used. I will investigate more in detail these two kinds of constraints, especially in unilateral commitment contexts (but also in joint commitments contexts since the Como types of cases raise problems). It will be the matter of the third part of this chapter.

Scientific liberty as absence of arbitrary constraint (domination)

Philip Pettit, who is partly interested in the same problem as Gilbert, notably the understanding of Durkheim's intuitions about the possible existence of groups or the relevance of contractualist models (Pettit 1993, 2003), has elaborated much more on this issue than Gilbert (Pettit 1999). And because Pettit considers "deliberative" situations (although not specifically in the scientific life contexts), his conceptions of freedom and liberty might help us to refine our ethical intuitions.[33] I will be led later on in this chapter to consider new historical cases, but in the meantime, I need to consider more general issues on freedom and liberty.

Pettit takes as his point of departure Isaiah Berlin's (1969) famous distinction between two kinds of liberty: positive and negative, and introduces a third kind. Pettit also introduces a new kind of Republicanism, based on this new kind of liberty, different from the classical Rousseauan tradition that arguably continues with Habermas and enhances positive liberty. *Positive liberty* is the freedom to decide politically for oneself, that is, to be politically autonomous. Rousseau's conception of contractualism is of this kind: citizens obey laws that they have decided on without delegates or any other intermediaries. Habermas's conception of deliberative democracy (Habermas 1992) just adds to Rousseau's conception the fundamental idea that what really makes the general will a value is the fact that it is obtained through deliberation, that is, through an exchange of arguments leading to a decision. If one accepts the idea that there may be deliberation in science, the construction of Durkheim and Mauss's programme of the sociology of categories seems close to this conception, because Durkheim and Mauss seem to have been jointly committed together (therefore, reciprocally and interdependently) without feeling any specific intellectual constraint from the other, as far as one can have a plausible knowledge of this kind of feeling on the basis of their confidences and of their general behaviour. And there is no evidence either that this was different regarding the other members.

But this Rousseau-Habermas normative conception of democracy seems pretty unrealistic for many reasons, of which the Como case is a good illustration. Thus, the Como case suggests that deliberation may be entirely lacking in joint commitment contexts and that the joint commitment may be coerced too. Moreover, as the relations between Bohr and his students in the last stages of the Copenhagen School or between von Mises and Machlup suggest, the joint commitment may be more apparent than real, so that what the junior scientist may have to face is actually a unilateral commitment associated with a constraint either to entirely submit to senior's thought or not to be allowed to participate in the modification of the shared principles.

The Rousseau-Habermas conception of democracy is still unrealistic for other reasons. Thus, as is well known, Berlin also *rejects* this conception of liberty, which he calls *positive liberty*, as have all the liberals since Benjamin Constant and Tocqueville, mainly because, according to them, it is the door open to the tyranny of the majority upon the minority.[34] These risks exist as well in the sciences and, for example, a majority may make remedy recommendations in medicine or a specific policy in economics even if the minority strongly disagrees. I will not elaborate more on these cases.

But given that all these reasons are more or less clearly accessible to everyone, it might not be surprising that when a new paradigm arises, scientists often *do not* jointly – nor unilaterally (if they can) – commit together, contrary to what Gilbert's model suggests. I will take here two examples, again chosen within the history of the research programmes I have already examined. In these two cases, while the scientists shared some fundamental ideas, they did *not* jointly commit and neither did they construct a collective research programme in Gilbert's sense of "collective". In these two cases, the collective beliefs or – rather – the collective acceptances were a matter of a "summative account" to use Gilbert's language. Besides, no scientists either unilaterally commit to one another. The issue will be, of course, to know whether this kind of "natural state" is both desirable and often realizable in science.

The first example is the case of the early Copenhagen School, during the few years *before* the Como conference and the Solvay Congress (which both took place in 1927). According to Mara Beller, and it is a point on which she insists a lot given the historical data at our disposal, although Bohr – not only the oldest of the group but the most famous as well at this period – Heisenberg, Pauli and Dirac shared certain if not very numerous conceptions on quantum mechanics, they *did not* commit to any collective programme (which does not necessarily exclude external institutional pressure).[35] In Berlin's and Pettit's terms, they experienced intellectual *negative liberty*, that is, *the independence from any* intellectual *constraint* (even from Niels Bohr), such as in a natural state (in the sense of Hobbes or Rousseau).

The comparison will be perhaps more tangible if I take the Austrian School of Economics also in its earlier period, that is, the period of Carl Menger, von Wieser and Böhm-Bawerk (of which von Mises was the most brilliant student). Like Bohr, Heisenberg, Pauli, Jordan and so on, they were *not at all* jointly committed to any collective programme although Menger, von Wieser and Böhm-Bawerk shared significant conceptions, particularly methodological individualism, subjectivism, focus on processes instead of equilibriums and also what is called the "marginalist" conception of value (Boettke 1994a). Like Bohr in the early Copenhagen School, Carl Menger was older than his colleagues and much better known. Nevertheless Menger, Böhm-Bawerk, and von Wieser remained independent of each other and Menger did not constrain at all their own projects (nor impose unilateral commitments).[36] So, they experienced negative liberty in Berlin's sense or freedom as *absence of interference* in Pettit's sense, that is absence of any (intellectual or institutional) constraint (Boettke 1994a).

Finally, Philip Pettit has introduced a *third conception* of freedom and liberty and has described a historical tradition that emphasises this conception, that is neither positive liberty in Rousseau's contractualist sense nor negative liberty or liberty as independence or absence of interference (and consequently any constraint) from others in one's project. This tradition, Pettit states, is closer to Cato's *Letters* or to the *Federalist Papers*. The key idea is that liberty is essentially the *absence of domination*, that is, absence of arbitrary constraint in a sense I will specify. This idea is directly opposed to the Hobbesian indifference to coercive commitments in both senses of coercion (coercion on the procedure and on the content of the commitment). It requires less than autonomy, that is, less than joint commitment and therefore less than reciprocal and interdependent commitments. On the other hand, it does not entirely exclude constraint as far as this constraint is not arbitrary (what Pettit calls interference), which means that *unilateral as well as joint commitment are not prohibited, if the required commitment does not mean loss of intellectual freedom* (what one can call the content condition) *and if the commitment is not constrained itself* (what can be called the procedural condition).

Where is the place of liberty in this conception of democracy? As in politics, Pettit states that liberty is not fundamentally located in the procedures of decision concerning the laws, in a scientific context, this is not fundamentally located in the formulation of the goals of the research programmes themselves, which can be formulated by an intellectual leader. But this liberty is situated in *the capability at least to publicly contest* the attempts, in politics, of the government – and, in a scientific context, those of the intellectual leader – to impose his view on the members of his team (e.g., see both the Heisenberg and Pauli case at Como, in a joint commitment context and the Weiszäcker case later, in a unilateral context).

One can furthermore argue, as James Bohman (2005) did following Hannah Arendt's intuitions, that this kind of conception is nevertheless slightly too weak and that the democratic minimum requires *the right and the capability of initiating modifications* (in law, in a collective research programme; see the Machlup case). I share Pettit's proposal and Bohman's amendment.[37]

Conclusion

To conclude, I will state the following propositions. First, the earliest periods both of the Copenhagen School of Quantum Mechanics and of the Austrian School of Economics are evidently characteristic of a scientific revolution, in which one can experience *negative liberty*. For this reason, these periods cannot probably be taken as references for further scientific development, which require collective planned research. Second, the French School of Sociology case seems a very rare case in which members seem to have experienced *positive liberty*, other cases of joint commitment or apparent joint commitment actually either leading to more or less moderate illegitimate intellectual coercion, as in the case of Bohr at Como, or hiding unilateral commitments leading to more severe cases of intellectual coercion, such as in the cases of the elder Bohr vis-à-vis Weizsäcker (and other junior scientists) and von Mises vis-à-vis Machlup.

The third model of liberty (*absence of domination*) seems more realistic in science. On the one hand, it does not require a complete absence of intellectual constraint (a certain degree of intellectual constraint might be useful for the scientific progress internal to a paradigm, that is when science becomes "normal science" in Kuhn's sense). It does not require joint commitment either: reciprocity and interdependence of commitments might be too demanding when there is an inequality of competence in a domain. But, on the other hand, this model of liberty nevertheless requires the right and the capability to publicly contest, which Heisenberg and Pauli in Como and Weizsäcker later on were lacking, and the right and the capability to initiate modifications in a research programme, which was missing for Machlup.

Notes

1. Regarding the place and the role of science in democratic states, which is another topic, see, for example, Kitcher (2001). And for the more general issue of ethical values in science, see, for example, Longino (1990, 2002).
2. See also Longino (1990, 2002) on a similar conception of social epistemology and ethics in science.
3. See Gilbert (2000).
4. I have not found any explicit references to Kuhn in Gilbert's work.
5. Kuhn himself insists only on passive aspects ("belief") of science, even during revolutionary periods, as if the change was due to a religious conversion (then from one "belief" to another "belief"). But this account of paradigm shift seems psychologically quite implausible, even if some passive processes can play a role, like new perceptual experiences. Besides, it is important to notice that the Copenhagen School is sometimes regarded as the *main* historical example on which Kuhn would have constructed the notion of paradigm (Hanson, 1958, Beller, 1999).
6. Gilbert does not use "acceptance" and "belief" in Cohen's technical sense, and she is even reluctant to this use, proposed by Meijers (1999) and Wray (2001). I do not

elaborate on this discussion, which does not really matter here. I just deem that Cohen's concepts clarify issues at stake here.

7. Nevertheless, one of my goals in this paper is to set forth the case that this model is still not realistic enough.

8. An example of an explicit collective manifesto, in the same historical context as the Copenhagen School, is the Vienna Circle manifesto, but this dealt with the most general philosophical bases of science rather than with a specific research programme in science. Another example is the first chapter of Bourbaki's *Éléments de mathématique*, but this case is not easy to investigate because the Bourbaki group cultivated so much secrecy about its functioning that even the identity of its members was hidden.

9. Unlike Thagard, Beatty (2006) explicitly uses Gilbert's concept of "joint acceptance" to account for similar collective medical recommendations (in this case, regarding the appreciation of genetics hazards of radiation exposure). Referring to Gutman and Thompson (2004), Beatty also tackles the issue of deliberation as a relevant decision-making procedure in such cases.

10. Gilbert would say that this is a *common knowledge* context.

11. Even explicit contracts may be more or less constrained, for example, when participants in a discussion do not dare to oppose a leader. But it is still more likely when the contract is implicit.

12. About the difference between deliberation and negotiation (or "bargaining"), see Elster (1991).

13. In fact, all the witnesses state that, while Durkheim always exerted enormous and almost totalitarian pressure upon Mauss (who was his nephew) to work more and more for the *Année Sociologique*, to the detriment of Mauss's private life, that was only (or at least mainly) *institutional* pressure or "coercion" (Besnard, 1979).

14. Gilbert (1996, 2006) shows convincingly how someone can jointly commit with other people who have previously jointly committed together.

15. "You will see a sample – maybe not as good as what you might have expected – of works of the French School of Sociology."

16. "In particular we have tackled the social history of categories of the human mind."

17. Dirac could not go to Como (nor could Schrödinger).

18. See also Bouvier, (2004 and 2007a). The case of Pauli is particular because Pauli participated very much in the rewriting of Bohr's paper between the Como talk and the various further publications, but not to the point of becoming a joint author with Bohr (like Durkheim and Mauss). The different versions of this paper were only signed by Bohr.

19. My account of the Bohr's Como lecture is based on the 1931 version, a revised version of the paper published in *Nature* in 1928, and the version to which scholars usually refer. The original oral version was not completely written down; consequently, it is not available (Mehra and Rechenberg 1982).

20. I thank Orly Shenker for this reference.

21. The main discussions at the 1927 Solvay Congress were held between Bohr and Einstein (who was very reluctant about Bohr's ideas on ontological indeterminism). Pauli participated very much in the discussion as well but there was no space for an internal discussion (or deliberation) within the Copenhagen group or within the slightly broader Copenhagen-Göttingen group. See Mehra and Rechenberg (1982) for further details.

22. Howard (2004) contends that what is now called the "Copenhagen interpretation" is actually very close to Heisenberg's conceptions (namely, the subjective

interpretation of quantum phenomena, focusing on the observer's role) and that it is a very regrettable mistake to attribute this conception to Bohr. Howard adds that Heisenberg presented himself explicitly as a spokesman of the whole Copenhagen group mainly from the fifties onwards (and that he might have specific personal interests in giving the impression that there was a unified interpretation shared by everyone in the Copenhagen group, and that he was really a member of this group, given that his very ambiguous behaviour during the Second World War regarding the Nazi nuclear programme had almost caused him to be banished from this group). The quotation drawn from Heisenberg and Born's report at Solvay reveals that it was already the case in 1927.

23. Similarly, one can argue that a consequence of the Solvay forced joint commitment is that sometimes Heisenberg's specific ideas (for example, the subjectivist interpretation of quantum mechanics) are attributed to Bohr, cf. the previous endnote.

24. Nevertheless most scholars think that the obscurity of this principle still grew as Bohr broadened its significance in his later publications.

25. See, for example, Rousseau (1964 [1762]) *Livre I*, chap. 4, 'De l'esclavage.'

26. Gilbert also discusses Simmons (1979) and Simmons (1984). See also Gilbert (2006).

27. Gilbert's analysis fits in with Weber's analysis of legitimation, which permits us to understand how (even authoritarian) political powers can stand up without necessarily using force or even threatening to use force.

28. The case of Durkheim, as described above, might be an exception.

29. Beller (1999) examines another similar case, Jesse Du Mond's (p. 274).

30. Weizsäcker later tried to unify science and religion. See Weizsäcker (1980 [1971]). The complementary principle's extensions provided by Bohr himself opened the door to such attempts.

31. There was also another similar occasion (regarding the right policies to encourage on the gold standard issue), in which there was a clash between Mises and Machlup; see Bouvier (2007b).

32. I addressed this specific issue in detail in Bouvier (forthcoming). See also Bouvier (2007b).

33. To my knowledge, Thagard, one of the leading figures of naturalized social epistemology, has not addressed the ethical aspect of cooperation or collaboration in science. See Thagard (1997).

34. Obviously, this tyranny is especially hard to support when the minority sincerely thinks that the majority is gravely mistaken.

35. Bohr tried to persuade Heisenberg in 1927, before the Como conference, *not* to publish his uncertainty paper as early as Heisenberg wanted. Heisenberg stated that "perhaps it was also a struggle about who did the whole thing first" (Beller 1999, pp. 13–39).

36. Unlike Heisenberg and Pauli, they already had a position as full professors.

37. Bohman introduces this idea in the context of a discussion of Habermas's (not Pettit's) ideas.

References

Beatty, J. (2006) 'Making disagreement among experts.' *Episteme* 3(1), 52–67.
Beller, M. (1999) *Quantum Dialogue. The Making of a Revolution*. Chicago: The University of Chicago Press.
Berlin, I. (1969) *Four Essays on Liberty*. Oxford: Oxford University Press.

Besnard, Ph. (1979) 'La formation de l'équipe de l' *Année sociologique.' Revue Française de Sociologie* 20, 7–31.

Boettke, P. J., ed. (1994a) *The Elgar Companion to Austrian Economics.* Aldershot: Edward Elgar.

Boettke, P. J. (1994b) Introduction to Boettke, *Elgar Companion,* 1–6.

Bohman, J. (2005) 'Is Democracy a Means to Global Justice? Human Rights and the Democratic Minimum.' *Ethics and International Affairs* 19(1), 101–16.

Born, M. and Werner Heisenberg (1927) 'Quantum Mechanics.' In: *Quantum Theory at the Crossroads: Reconsidering the 1927 Solvay Conference,* ed. G. Bacciagaluppi and A. Valentini (2006), Cambridge: Cambridge University Press, pp. 408–47.

Bouvier, A. (2004) 'Individual Beliefs and Collective Beliefs in Sciences and Philosophy. The Plural Subject and the Polyphonic Subject Accounts. Case Studies.' *Philosophy of the Social Sciences* 34(3), 382–407.

Bouvier, A. (2007a) 'Collective Belief, Acceptance and Commitment in Science. The Copenhagen School Example.' *Iyuun: The Jerusalem Philosophical Quarterly* 56, 91–118.

Bouvier, A. (2007b) 'Qu'est-ce qu'un engagement de groupe en sciences sociales? L'exemple de l'école autrichienne d'économie.' In: *L'épistémologie sociale. Une théorie sociale de la connaissance,* ed. A. Bouvier and B. Conein (2007) Paris: Ed. EHESS/ Raisons Pratiques, pp. 255–94.

Bouvier, A. (forthcoming) 'Passive Consensus and Active Commitments in the Sciences.' *Philosophy of Science Association Congress,* 6–8 November 2008, Pittsburgh.

Caldwell B. (2004) *Hayek's Challenge. An Intellectual Biography of F. A. Hayek,* Chicago: University of Chicago Press.

Cohen, L. J. (1992) *An Essay on Belief and Acceptance.* Oxford: Clarendon Press.

Durkheim, E. and M. Mauss (1903) 'De quelques formes primitives de classification: contribution à l'étude des représentations collectives.' *L'Année sociologique* 6, 1–79.

Elster, J. (1991) 'Arguing and Bargaining in the Federal Convention and the Assemblee Constituante.' *Working Paper.* Occasional papers from the Law School, the University of Chicago.

Gilbert, M. (1989) *On Social Facts.* Princeton, NJ: Princeton University Press.

Gilbert, M. (1993) 'Agreements, Coercion and Obligation.' *Ethics* (103), reprinted in Gilbert 1996, chap. 12, pp. 281–311.

Gilbert, M. (1994) 'Durkheim and Social Facts.' In: *Debating Durkheim,* ed. W. Pickering and H. Martins. London, Routledge.

Gilbert, M. (1996) *Living Together. Rationality, Sociality and Obligation.* Lanham, MD: Rowman & Littlefield.

Gilbert, M. (2000) 'Collective Belief and Scientific Change.' In: *Sociality and Responsibility,* ed. M. Gilbert. Lanham, MD: Rowman & Littlefield, chap. 3.

Gilbert, M. (2002) 'Belief and Acceptance as Features of Group.' *Protosociology* 16, 35–69.

Gilbert, M. (2006) *A Theory of Political Obligation. Membership, Commitment and the Bond of Obligation.* Oxford: Oxford University Press.

Goldman, A. (1999) *Knowledge in a Social World.* Oxford: Clarendon Press.

Goldman, A. (2002) *Pathways of Knowledge.* Oxford: Oxford University Press.

Gutman, A. and D. Thompson (2004) *Why Deliberative Democracy?* Princeton, NJ: Princeton University Press.

Habermas, J. (1992) 'Further Reflections on the Public Sphere.' In: *Habermas and the Public Sphere,* ed. C. Calhoun. Cambridge MA: MIT Press.

Hanson, N. R. (1958) *Patterns of Discovery.* Cambridge: Cambridge University Press.

Howard, D. (2004) 'Who Invented the "Copenhagen Interpretation": A Study in Mythology.' *Philosophy of Science* (71), 669–82.

Kitcher, Ph. (2001) *Science, Truth, and Democracy.* Oxford, Oxford University Press.

Longino H. (1990) *Science as Social Knowledge*, Princeton, NJ: Princeton University Press.

Longino, H. (2002) *The Fate of Knowledge.* Princeton, NJ: Princeton University Press.

Machlup F. (1981) *Homage to Mises.* Hilldale, MI: Hillsdale College, pp. 19–27.

Mauss, M. (1950) *Sociologie et anthropologie.* Paris: Presses Universitaires de France.

Meijers, A. (1999) 'Believing and Accepting as a Group.' In: *Cognition and the Will*, ed. A. Meijers. Tilburg, The Netherlands: Tilburg University Press, pp. 59–71.

Mehra, J. and H. Rechenberg (1982) *The Historical Development of Quantum Theory.* New York: Springler-Verlag.

Merton, R. (1942) 'Science and Technology in a Democratic Order.' *Journal of Legal and Political Sociology*, T.1, 115–26.

Mises (von), L. (1966) *Human Action.* Chicago: Contemporary Books.

Pettit, Ph. (1993) *The Common Mind. An Essay on Psychology, Society and Politics.* Oxford: Oxford University Press.

Pettit, Ph. (1999) *Republicanism. A Theory of Freedom and Government*, 2nd ed. Oxford: Oxford University Press.

Pettit, Ph. (2003) 'Groups with Minds of their Own.' *Philip Pettits's Papers*, Philip Pettit's Homepage, Princeton University.

Rizzo, M. (1990) 'Mises and Lakatos. A Reformulation of Austrian Methodology.' In: Littlechild St. (1990) *Austrian Economics*, vol.1. Aldershot: Edward Elgar, pp. 487–507.

Rousseau, J.-J. (1964 [1762]) *Contrat Social in Rousseau Œuvres complètes*, T. III. Paris: Gallimard, Pléiade.

Schmidtz, D. (1990) 'Justifying the State.' *Ethics* 101(1), 89–102.

Simmons, A. J. (1979) *Moral Principles and Political Obligations.* Princeton, NJ: Princeton University Press.

Simmons, A. J. (1984) 'Consent, Free Choice, and Democratic Government.' *Georgia Law Review* 18, 809–17.

Sztompka, P. (2007) 'Trust in Science.' *Journal of Classical Sociology* 7(2), 211–20.

Thagard, P. (1997) 'Collaborative Knowledge.' *Noûs* 31(2), 242–61.

Thagard, P. (1998a) 'Ulcers and Bacteria I: Discovery and Acceptance.' *Studies in History and Philosophy of Science* 29, 107–36.

Thagard, P. (1998b) 'Ulcers and Bacteria II: Instruments, Experiments, and Social Interactions.' *Studies in History and Philosophy of Science* 29, 317–42.

Wray, K. B. (2001) 'Collective Belief and Acceptance.' *Synthese* 129, 319–33.

Weizsäcker, C. F. (1980 [1971]) *The Unity of Nature.* New York: Farrar Straus & Giroux.

Part IV

The Democratic Governance of Social Science

8
Public Sociology and Democratic Theory

Stephen P. Turner

On the eve of the Second World War, the Danish sociologist Svend Ranulf wrote a *samizdat* study in the history of sociology, which he later published under the title 'Scholarly Forerunners of Fascism' (1939). It is a stark reminder that sociology has always been in conflict or potential conflict with liberal democracy. Although classics such as those by Émile Durkheim and Ferdinand Tönnies were also prominently featured, August Comte earned a privileged place in Ranulf's text for his attacks on the idea of freedom of conscience and the 'anarchy of opinions' in his time. These make for shocking reading even today. Comte denounced the very idea of a liberal public making decisions based on discussion, ridiculed citizens for forming their own opinions, and called for state control over the dispersion of ideas.

> There is no liberty of conscience in astronomy, in physics, or even in physiology, that is to say, everybody would find it absurd not to have confidence in the principles established by men who are competent in these sciences. If it is not so in politics, this is due only to the fact that, since the older principles have yet to be abandoned, and no new ones have as yet been devised to replace them, there are really no fixed principles at all in the meantime. (Comte [1830–42] 1864, IV: 44n)

The argument is compelling, as even John Stuart Mill, who broke with Comte over it, conceded (1865). Comte also argued that the lessons of sociology should be imposed on the population through a regime of indoctrination (Comte [1830–42] 1864, IV: 22, 480; V: 231). If sociology was taken to be, as Comte took it to be, a science that had, for the practical purpose of ordering political life, reached the positive stage, it ought to be taught as scientific truth, not as a matter of opinion. To object to teaching it in this way would be similar to objecting to teaching evolution as fact: a kind of willful obscurantism.

Why did Ranulf regard this reasoning as fascistic? Writing under the Nazis, he was sensitive to one highly overt kind of intervention by the state: indoctrination. The idea that the state should be in the business of indoctrinating students, as Ranulf saw, was continuous with totalitarianism. Indeed, the following might be a good definition of totalitarianism: for the totalitarian regime, all significant matters that were formerly thought to be matters of opinion are placed in the category of truths about which there can be no legitimate dispute and which are thus appropriately the subject of indoctrination. Ranulf of course disagreed with Comte's assessment of the state of sociology, and specifically with the idea that it had attained such a level of development that it could be regarded, in practice, as truth rather than opinion.

But the issue cannot be regarded as having died with Comte, or even as being closed. Instead, it points to a pervasive issue within democratic theory. What does it mean for liberal democracy if a significant part of public discourse is false or erroneous? Democratic theory assumes that this discourse is self-regulating and that error will be corrected by discourse itself. But what if there is genuine expert knowledge about the kinds of things that public discourse takes as its object, but public discourse does not acknowledge it? What if the errors cannot be readily corrected within public discourse itself, precisely because understanding them requires expertise beyond that of the ordinary citizen? What if the terms of public discussion, or the conditions of its conduct, are such that the truth on certain issues cannot come to the fore?

Nor is this a purely hypothetical possibility. Elements of the public routinely dispute or reject academic claims, including those of natural science. Does this discredit the basic ideas of democratic theory? Comte's answer would of course have been yes. The question of what to do about the ignorance and incompetence of the public is a draconian one: abolish public discussion in favor of expert rule and indoctrination. And there is a sense in which this was tried by Communism (Turner, 2006). But what is the proper democratic response?

It should be evident that this question is closely connected to the problem of public sociology. Many of the most interesting comments on the issue have addressed precisely this kind of problem (e.g., Stacey, 2004). Issues of political theory have for the most part been ignored in the discussion of public sociology. However, I will argue, Burawoy's proposal (2005a) needs to be understood in terms of the larger problem of finding a place for sociology in liberal democracy. In this respect, it will become evident that Burawoy's proposal is more significant than merely calling for sociologists to better communicate with the public. It represents a novel answer to a whole range of questions about the political meaning of sociology. Moreover, I will suggest, Burawoy's discussion is not merely a proposal. It is a radically new understanding of the kind of discipline sociology has actually become.

The problem of public opinion

Liberal democratic theory is absolutist about the political neutrality of the state. 'The conception that government should be guided by majority opinion makes sense', as Hayek puts it,

> Only if that opinion is independent of government. The ideal of democracy rests on the belief that the view which will direct government emerges from an independent and spontaneous process. It requires, therefore, the existence of a large sphere independent of majority control in which the opinions of the individuals are formed. (1960: 109)

This is the meaning of the political neutrality of the state. For the state to intervene and shape opinion, for example by indoctrination, makes the 'democratic' subordination of the state to public opinion into a sham by making the public process of discussion into a sham. This basic idea can also be understood as involving fundamental rights. Jefferson put it thus: 'To compel a man to furnish contributions of money for the propagation of opinions which he disbelieves and abhors is sinful and tyrannical' (1779, Papers 2:545). The background to Jefferson's claim is the issue of state-sponsored religion or 'establishment'. But the form of the claim is to affirm a right against a certain kind of state coercion, coercion to support the propagation of opinions one rejects.

The fact that both of these quotations rely on the term 'opinion' is critical to understanding the issue: the considerations of principle are absolutes. Either the state, the source of compulsion, intervenes in the discussion that makes democracy meaningful or it does not. But the question remains: who decides what is fact and what is opinion, and how is it decided? If the state decides what the line is, against the opinions of some of those it governs, it becomes a propagator of a second-order doctrine of the distinction between fact and opinion. And if this is itself subject to dispute, people are deprived both of their right not to support the propagation of beliefs they reject and of the right of democratic citizenship to rule by opinion that is formed independently and spontaneously.

These basic liberal premises present a dilemma. If sociology is opinion, that is to say, if we put it into the same category as religious dogma, citizens ought not to be forced to subsidize it or have it directed at them. It is no more than state-sponsored propaganda. Alternatively, if sociology is a science, a science of facts, we may grant it the kind of exemption from this restriction that natural science receives. Sociology can be part of the educational process and promoted by the state, if it is neutral. If it is a science that can have its full effect only by being accepted as true, the inculcation that offended Ranulf would be as appropriate as a public health campaign. Sociologists who failed to act on their knowledge by giving it to the public would be morally culpable. And states that failed to make use of it would have failed to act on behalf of its citizens.

Yet drawing the line between fact and opinion, rather than resolving this dilemma, leads to endless muddles here and elsewhere. Paul Feyerabend argued in *Science in a Free Society* that science itself was one opinion among many and, thus, should not be imposed by the state as dogma, as it is through educational institutions (1978: 91–2, 106–7). But radically expanding the notion of opinion leads to absurdities, or, more importantly, what appear in given political traditions, to be absurdities. Should the state not teach science at all? If it attempts to be nondogmatic in the teaching of science, does that mean that such things as creation science need to be taught?

Almost no one regards these conclusions as reasonable. But convincing justifications in principle in response to such questions are not readily available. Grounding the distinction in the philosophy of science has not proved satisfactory: there are too many different accounts of the boundaries of science to think that any of them can be treated as fact. The very existence of diversity of opinion here poses a problem: unless we think that philosophies of science can be taken to be fact rather than opinion (something assumed, rather strangely, by U.S. Court decisions in cases involving creation science). Choosing one over another to draw the line is a matter of choosing one opinion over another, and is thus a violation of neutrality. It also raises a political question: who decides? In an important sense, Carl Schmitt's dictum that what is 'unpolitical' is a political decision ([1922] 1985:2) is true, even for, and perhaps especially for, liberal democracies.

The distinction between fact and opinion that is central to the liberal principle is controversial even within liberalism. Many liberal thinkers agree, tacitly or explicitly, that it is a mistake to draw the line between science and other forms of opinion in an absolute way and argue for a different way of understanding the relation of science and politics (cf. Polanyi, [1946] 1964; Turner, 2005). But in practice, matters are more complex. Some decisions to draw the line between opinion and fact are overtly political: the ancient Greeks regarded the oracle at Delphi as, in effect, an impartial expert, putting its pronouncements in the realm of fact and treating them as such as part of decision-making processes. Decisions like those in the American courts to define creation science as religious are political in the sense that they are an expression of the authority of the courts, a political body. But from the point of view of the courts themselves, they are not and cannot be: courts are politically neutral. Moreover, disciplines make their own claims to expertise, often involving rules that they enforce, directly or indirectly, on their members and that they seek to have recognized, for example, as a profession or university subject. So do individuals who make the case in the court of public opinion for their own claims to be accepted as fact.

Some decisions are conventional or part of national traditions. The particular problems of church and state that arise in connection with the teaching of evolution, for example, are the products of a long-running American legal issue: in Europe, the conventional lines are drawn differently, as such issues as the dispute over Danish blasphemy against Islam shows. French *secularité* requires, or so it has seemed to many, secularity in the dress of schoolchildren, something that would fall into the category of religious freedom in the United States.

There is another critical line-drawing problem that has been alluded to here. Indoctrination is an extreme and overt form of state action. But there are many other ways in which the state intervenes in public discussion. Hayek used the phrase 'independent and spontaneous' to describe the process of democratic discussion. But what does this mean? A traditional view holds that it is sufficient for the *state* to be politically neutral. But what if, as Habermas argued ([1962] 1989), the social structure that underlies 'public' discussion is itself biased toward certain systematic errors and illusions? What if the process is distorted by non-\governmental factors or by indirect consequences of state actions? For example, if there are multiple sides to a controversy, what if some sides have access to resources that allow their views to be made persuasive beyond what is rationally justified? Does the state have an obligation to support corrections to this process, such as attempts to balance the discussion? Or to intervene with research to improve its quality?

The notion of indoctrination raises questions of its own. What is the line between the propagation of opinions or the intervention in democratic discussion – discursive will-formation, to use the Habermasian term – and something that is not proscribed, such as, education? Romanticist critics of the enlightenment denied that there was such a distinction, and today it is an antiliberal commonplace to regard schoolbooks and the schooling experience as sources of sexism, racism, and the reproduction of class distinction (Bourdieu, 1984; Bourdieu and Passeron, 1990; Bourdieu, Passerson, and Sainte-Martin, 1994).

Thinkers on the libertarian end of the spectrum like Hayek tend to regard the issue narrowly. For them, the public discussion we have is good enough. The primary risk is in the intrusion of the state. But such thinkers also take the category of proscribed state intervention very broadly. For thinkers on the democratic Left like Habermas, the promise of liberalism is chronically unfulfilled for reasons that arise from society rather than the state. For them, without the intrusion of the state in society, genuine discussion, discussion in which the argument with the greatest rational force wins, will remain out of reach. So for them, state intervention and regulation are not only justified but also necessitated by what we might think of as the market failures of the marketplace of ideas (Dahl, 2006; Pettit, 1997: 194).

The practice of political neutrality

These are issues in theory, but theory sometimes informs political practice in consequential ways. In practice, the range of 'liberal' solutions to these problems has been large. The idea of subsidizing science (or for that matter religious groups as self-governed communities) can be regarded as consistent with liberalism if this is done in some fair procedural manner, and is thus not partial to any particular opinion. State support of universities is paradigmatic of this: they are 'autonomous' associations, which are nevertheless expected to respect their own limited purposes. In theory, there is no issue with private institutions, but ordinarily the state has a role in certifying institutions and recognizing their graduates. Because contemporary sociology is largely centered in universities, my focus in what follows will be on this institutional form. But it should be noted that much of the history of sociology involved private citizens or private funding, and much 'public' sociology involved private citizens writing to support themselves.

Liberal democracies have steered a complex middle course in the face of these issues by encouraging, subsidizing, and granting partial autonomy, a separation from direct political involvement and control, to special institutional structures that support expertise by education and research.[1] Foundations, for example, operate under a 'contract' that allows them to avoid taxes as long as they spend money for charitable purposes: politics in the explicit sense of partisan electoral politics are excluded. But policy studies, advocacy of various kinds, as long as it is not direct advocacy of pending legislation, and so forth, are allowed, along with traditional charitable activities. What is the rationale for this arrangement? The basic idea is this: it is better for the rich to give money to charity than to spend it on luxuries or on their heirs, so the state permits the creation of foundations with the incentive of no taxation. In return, the foundations must comply with the law or they face the possibility of having their tax-free status revoked.[2]

The political logic behind these relationships is rarely discussed. There is a large literature on science and its political meaning. Michael Polanyi, for example, who argued, like Feyerabend, that science was a form of opinion, also argued that it deserved its exemption from the restrictions of liberalism because science was an autonomous, self-governed community with its own core commitments and purposes that shared features of liberalism, served society at least indirectly, and needed to be treated like an established church or the legal profession ([1946] 1964:67). A similar argument was made about science by Don Price in the United States, likening science to an 'estate' (1965). Sociologists are most familiar with what was, at the time it was published, a minor contribution to this discussion: Merton's paper 'A Note on Science and Democracy' (1942), better known by one of its later titles, 'The Normative Structure of Science' in *The Sociology of Science* (1973).[3]

The contract with science was this: fund science and leave it alone to determine its research priorities; in return, science would yield results of significant practical value for social needs. This was an instrumental contract. Neutrality was dealt with implicitly by virtue of the fact that science, meaning natural science, had its own strong tradition and a sense of its boundaries. Merton called this an ethos and described it in terms of communism, universalism, disinterestedness, and organized skepticism. The autonomy of science, he argued, was justified by the fact that scientists policed themselves in accordance with this ethos.

One of Merton's concerns in this paper was with the autonomy of sociology, which he recognized was threatened by the fact that sociological claims often conflicted with popular beliefs. The concern proved to be well founded. Sociology for the most part has been excluded from this contract, in part for reasons that relate to the problem of neutrality. A significant group of scientists opposed the inclusion of the social sciences into the National Science Foundation, in part on the grounds that the social sciences were political or ideological rather than genuine sciences. They succeeded in excluding the social sciences from the original legislation. When the foundation eventually relented, it did so with the understanding, enforced by the choice of program officers and by their decisions, that the sociology supported by the foundation would be quantitative and apolitical. But sociology had its own version of the contract, and its history is essential to understanding the case for 'public sociology'. The implicit deal that has been the basis of sociology's claim on society has been that supporting sociology as a research and academic discipline will produce results that have a positive effect on societal problems.

At many points in the past, this deal has been articulated explicitly. The vast investment of Rockefeller funds in social science in the 1920s was understood as a means of improving the quality and capacity for research, which in a few years was expected to be applied to and to have an impact on social problems. When this did not occur, funding was cut (Turner and Turner, 1990: 41–5). The experience of the Rockefeller philanthropies was followed by that of the Ford Foundation, which was presented with the promise that an investment in social and behavioral science would provide the tools to deal with issues of race, which were understood as attitudinal problems. Needless to say, the issues of race did not vanish in the face of new social psychological methods; attitude theory itself was revealed as intellectually empty. Ford money was soon directed elsewhere. More generally, the investment in pure sociological research in the narrow sense, and indeed in academic sociology generally, has never had significant positive results for social problems. But the same consideration of impact on social problems was an explicit part of arguments for funding sociology long after, for example, the attempt by Senator Fred Harris in the 1960s to create a National Social Science Foundation.

Sociology never depended on this contract for its primary income. That came from teaching. As an academic subject, sociology operates under the autonomy granted to the university and to such bodies as foundations. In the university, fields that conflict with public opinion can survive, even flourish, by attracting students who are skeptical of public orthodoxies. And at various times sociology has flourished, precisely in its role as an alternative to public opinion. But as with foundations, the autonomy of universities is granted under certain conditions, one of which is that they remain outside of politics. In the course of establishing itself as an academic subject, sociology had to draw its own lines in a way that was sufficiently convincing that other scholarly fields accepted it as a legitimate discipline. And the admission of sociology into the university provided a kind of resolution to the problem of neutrality. Within these institutions, the issue is transformed, but does not vanish. It is often transferred to the specific details of the institutional arrangements themselves. But the logic of political neutrality does not become irrelevant. There are still boundaries, such as the limits of the tradition of academic freedom, and issues like 'is it education or indoctrination' still arise.

Sociology has, throughout its history, flirted with the boundaries set by these relationships and has a long internal history of discussing them. A significant part of the background to the current discussion of 'public' sociology is the fact that a discussion has opened up in the United States over the politicization of the universities (Horowitz, 2006). It is routinely alleged that academics have stepped over the line by producing biased scholarship and teaching and by punishing students who present views in class that are acceptable, even conventional, in the larger body of public discussion, but are treated as false and even excoriated as racist, sexist, and so forth by classroom teachers. This new discussion of professorial political bias is particularly relevant to Berkeley sociology and its form of 'public sociology'. The major critic of politically motivated abuses of the academic status, David Horowitz, was an associate of some of the present Berkeley faculty in the 1960s when they were all involved in radical politics. Horowitz became an apostate and critic of politicization (1998); Burawoy a defender of public sociology.

Burawoy's new model

The implications of the discussion so far are straightforward enough. There is a conflict in the abstract, recognized explicitly in Comte and others, between liberal democracy and a 'social science' that makes 'political' pronouncements, whether it does so by asserting intellectual authority over topics that public discussion takes as its domain or by participating as a state-sponsored source of opinion within public discussion. This conflict takes other forms as well, such as the issue of classroom indoctrination. How does the literature on public sociology handle this conflict? And how does liberal democracy handle sociology?

Traditionally, democratic theory has handled questions about expert knowledge by defining them away. In Rawls, for example, public reason is contrasted with private reason. Public reason is treated as the last word on issues in the public realm, such as matters of the purposes of the state, because there is no authority higher than the result of public discussion; science, in contrast, is a form of private reason, which has no special authority on these questions (cf. Rawls, 1993). But while this distinction may work for physics – no one supposes that debates on elementary particles ought to be conducted by parliaments – it does not work very well for the social sciences, which often make claims of the same kind as politicians.

Thus, sociology occupies an uncomfortable and anomalous position. Sociology purports to have expert knowledge about matters that are in the domain of public discussion in liberal democracy. Its value lies in the ability of sociology to do something that public discourse ordinarily cannot or does not do. The claim that sociology makes for public support and recognition depends on this purported ability. But once recognized and supported, what status do sociological claims have? Are these claims merely another contribution to the debates that make up 'public reason'? Or are they properly understood as claims that serve to take issues out of the realm of public discussion and into the category of fact or expertise proper?

How one answers this question has implications for the question of funding. Ordinarily, the state in liberal democracies does not fund political viewpoints. If sociological claims are part of public reason, they would run afoul of this practice. But if sociology were a more limited activity, it would undercut the *raison d'être* for sociology, for it would limit its contribution to public discussion. The issue can be put differently. Is sociology, as a publicly funded activity, an anomaly for liberal democracy, fundamentally in conflict with it because it is an attempt to usurp the functions of public discussion by expertizing it? Or is it a means by which liberal democracy is supported and improved and a legitimate object of state support?

It is in his approach to this question that Burawoy comes to a genuinely novel and radical conclusion. But the argument is not made directly. The literature on public sociology begins with a few core ideas: that sociology has gone Left while the world has gone Right, as Burawoy puts it (2005a: 261); that sociology has a direct contribution to make to democratic discussion that is frustrated by various factors; that engaging the public, especially by critiques of conventional and especially right-wing views, is a moral obligation of sociologists; that sociologists are disadvantaged in the realm of public debate by the reductive and simplistic character of public discussion (Stacey, 2004) and the lack of respect for sociology, and that other fields have greatly outpaced sociology in the 'public intellectual' market, to the detriment of the field.

Burawoy's explicit contribution to this literature is that he has outlined a model of public sociology and sketched out the relations between various forms of present sociology, particularly the relation between activist public sociology and what he calls professional sociology. His innovative concept in the key paper is the idea of an organic relation between sociology and its various publics (2005a: 263–66).[4] The examples he has in mind include sociologists' participation in nongovernmental organizations (NGOs) and social movements, a relation that he suggests is dialogical rather than expert. This is, superficially, a 'liberal' idea; Comte, as we have seen, would have nothing of dialogue.

Burawoy, however, has a novel approach to the problem of neutrality. In addressing this problem, Burawoy is implicitly acknowledging that sociology as an ideological activity – as a discipline with its own value scheme – would be a serious problem for liberal democracy. If 'public sociology' was, to put it bluntly, an ideology or set of ideologies with some supporting factual content, subsidizing it would violate the political neutrality of the state, represent an intervention in public discourse on a particular side, and oppress by virtue of coercing citizens to pay for themselves to be subjected to propaganda. This is a trap that arguments for a politicized sociology of the 'whose side are we on?' form routinely fall into.

Burawoy, however, is subtler than this. Rather than rejecting the idea of neutrality outright, he appeals to the idea of neutrality. In a crucial passage, he claims that there is 'no intrinsic normative valence' to the idea of public sociology (2005a:266). Without fully articulating it, he nevertheless indicates a model of the relation to values to public sociology: the character of particular expressions of public sociology – books, relations of support, and cooperation with NGOs, and the like – reflect the individual values of sociologists.[5] But these are taken to be private matters, as with Comte, and incidentally American sociologists like Ward, rather than as matters that sociology itself can decide.

When sociologists ally themselves to a social movement in an organic manner, they do so out of individual choice. They would still be in the bounds of public sociology even if the movement in question were fundamentalist Christianity. This means that the idea of a public sociology is not intrinsically Leftist, despite the fact that most sociologists are on the Left. Moreover, it means that sociology as a discipline is not responsible for the 'public sociology' commitments and alliances that individual sociologists undertake in the name of sociology and in accordance with their own conscience. We can, in short, recognize a work as a piece of good public sociology while rejecting the values on which it is based.

The legitimacy of sociology, for Burawoy, depends in part on the successes of professional sociology in providing rigorous social research. But it also depends on the dialogic success of public sociologists' organic relations to social movements and to the audiences, especially student audiences, which sociology cultivates a relation to. These relations also provide an important kind of constraint on sociologists: they are compelled to express themselves in ways that fit with and articulate the experiences and convictions of these movements. Thus, the sociologist in an organic relationship is engaged in the task of providing the self-understanding of society, in the phrase of Edward Shils, but no longer for society as a whole, but for a society made up of different standpoints.

The meaning of the argument

If we read Burawoy, as I think we should, as not so much advocating but rather ratifying the new relation of sociology to the political, the significance of the argument becomes clearer. American sociology, of which Berkeley sociology is an extreme case, is now primarily sustained by relationships of the sort he describes. The model is women's studies, which purports to represent the standpoint of women, and stands in a special relationship of mutuality, organic and dialogic, with the feminist movement. Like women's studies, the appeal of sociology to students is bound up with a kind of self-understanding that involves identities, especially those of race and gender. The actual scholarship of sociologists is primarily motivated by public issues raised by such movements. Theory is instrumental: cultural sociology, Foucault, Bourdieu, and the like, are particularly useful in a sociology that provides a self-understanding with an emphasis on resistance and victimization.

Burawoy's achievement is to provide an articulation of the normative content of the social relationships that sustain this kind of sociology – its contract with society. In this respect, it represents a significantly novel approach to the problem outlined above: finding a *modus vivendi* with liberal democracy. As such, it belongs alongside such works as Weber's *Wissenschaft als Beruf* ([1917] 1946) and his essay on value-neutrality ([1917] 1949), with which it sharply contrasts. And its implications, many of which are not stated by Burawoy, are very far-reaching.

The central political implication involves the place of sociology in relation to liberal democracy. The kind of sociological scholarship Burawoy is legitimating under the heading of organic public sociology is advocacy scholarship: it consciously attempts to understand and articulate the standpoint of some group in society in a way that it intelligible to and instructive to the members of the group, as well as to support this standpoint through social research. The term 'dialogical', whose ordinary contrastive is 'analytic', implies in this context that the concerns and viewpoints of the partners are not merely topics for research but enter into the definition of the problem. It may be noted that this use of dialogue is not liberal: the partner in the dialogue is not the public as a whole, but the movement allies.

Under previous models of sociology and its role in liberal democracy, this kind of scholarship would be a problem both for sociology and for the state: for the state, because it represented a form of taking sides; it would be a problem for sociology, understood as a science, because it lacked what Merton called disinterestedness. Burawoy's model, if we understand it in the larger context of democratic theory, allows sociology to transcend both problems. Many liberal democracies subsidize viewpoints under some form of the neutrality assumption. Burawoy supplies a parallel neutrality assumption: group advocacy scholarship, or what he characterizes as audience-specific organic public sociology, is acceptable in principle as public sociology regardless of the group.

This kind of scholarship is not impermissible state subsidization of viewpoints, but rather a means of improving the quality of public discussion through the subsidization of opinion diversity. This transcends the problem of political neutrality because the state does not chose, and the state-subsidized discipline of sociology itself is neutral between the kinds of commitments that individual sociologists choose to make. Sociology is understood to be one of many participants in the larger public debates in which a flourishing liberal democracy engages. It is understood as well that there are viewpoints of other, contrary kinds, which are also subsidized and also contribute to competition in the forum of liberal democratic public discussion.

The problem of intellectual neutrality or disinterestedness is solved in a different way, and it is crucial to see what the full implications of this solution are. Burawoy appeals to the standpoint epistemology of Patricia Hill Collins ([1990] 2000) and Dorothy Smith (1993), which is rooted in a larger feminist epistemological literature. Weber had argued that values entered into the description of social phenomenon, but not into the casual analysis done using these valuative descriptive categories. Gunnar Myrdal put this same point differently in the methodological appendix to *An American Dilemma*: sociology can say something scientifically about the correctness of people's beliefs, but nothing about the correctness of their valuations ([1944] 1962:1027–34; cf. Eliaeson, 2006). Standpoint theory draws the line differently: beliefs *and* values are determined by one's standpoint, and evidence enters into a standpoint – facts become facts – according to standards specific to that standpoint. One variant of standpoint theory holds that certain standpoints, those of the oppressed, are epistemically privileged, because privilege blinds its possessors to truths about privilege. Another variant holds that standpoints are irreducibly plural and in conflict. Disinterestedness, in either case, is not possible.[6]

Yet public sociology in the organic mode, despite its engaged or standpoint character, may be said to be legitimately part of academic scholarship and appropriately supported by the state if it serves the purpose of improving democratic discussion. The advocacy studies that are done by sociologists organically related to social movements are valuable only if they are persuasive, and to be persuasive they need to be up to professional standards. But adhering to these standards is merely a means of gaining public credibility – legitimacy, as Burawoy puts it (2005a:267). They do not establish the objective truth of the conclusions in any sense similar to that of 'positivist' science. The conclusions are rather accredited by professional consensus as claims to be taken seriously in public discussion. This is the end point and aim of this kind of sociology. The political meaning of sociology is thus to contribute to the diversity of political discussion by helping to give voice and support to particular movements and groups.

Epilog

The state preserves its neutrality by virtue of the diversity of the standpoints it supports. I began this chapter with the Comtean idea of the intellectual authority of scientific sociology supplanting the anarchy of opinions of liberal democratic discussion. The discussion of Burawoy closes with the idea of organic public sociology as a handmaiden to partisan participants in liberal democratic discussion. The transformation is profound. Comte and those after him (and for much of the twentieth century), the Left and reformers (such as Karl Pearson and the Webbs), looked to sociology to solve the problem of the anarchy of opinions. Where Comte believed politics resolved nothing, Burawoy believes that science resolves nothing – that standpoints are irreducible. Past sociologists, such as Merton, defended sociology in its conflicts with public opinion by stressing its self-imposed restrictions, such as universalism and disinterestedness; Burawoy seeks to justify sociology for its contribution to, and intimacy with, particular social movements and causes in their political efforts.

The former goal of sociology was to get a seat at the table of the sciences; the goal of public sociology is to kibitz at the table of political discussion. Although Burawoy does not formally abandon the scientistic aspirations of sociology, he pointedly reminds us that the reign of scientism was brief, partial, and unsuccessful (2005a:17–19) and that sociology as a discipline recovered its lost student audience at the same time that the scientistic dream of a coherent, cumulative science evaporated. This is a conception of sociology that thus has the virtue of realism about the actual prospects of sociology as a viable discipline, which the scientizers never did, as well as solving a problem that the advocates of politically committed sociology, such as Lynd, failed to solve: the question of how a neutral state could subsidize it.[7]

Notes

1. This problem was the focus of Merton's 'Norms' essay (1942; [1949, 1957] 1968; Turner, 2007a), but formulated as a problem in political theory by Michael Polanyi (cf. Polanyi 1939, 1941–43, 1943–45, [1946]1964, [1951] 1980).
2. The terms of this 'contract', however, go beyond the law, which can be revoked or changed at any time. It is revealed in the various public investigations into foundations that have occurred when politicians believed the foundations had strayed into politics. From the point of view of liberal democratic theory, these investigations, and the ensuing threats to change the law if the foundations didn't

stay out of politics, served to mark the boundaries of acceptable behavior (Nilsen, 1972:53–4).
3. The history of the complex issue of the place of science in relation to liberal discussion is discussed at length elsewhere (Turner, 2007b).
4. The term derives from Gramsci's notion of the organic intellectual (1971).
5. Ironically, the value theory that underlies this way of making the distinction is Weberian.
6. See Susan Hekman (1997) 'Truth and Method: Feminist Standpoint Theory Revisited' and 'Reply to Hartsock, Collins, Harding, and Smith'; Patricia Hill Collins (1997) 'Comment on Hekman's "Truth and Method: Feminist Standpoint Theory Revisited": Where's the Power?'; Sandra Harding (1997), 'Comment on Hekman's "Truth and Method: Feminist Standpoint Theory Revisited": Whose Standpoint Needs the Regimes of Truth and Reality?'; Nancy Hartsock (1997) 'Comment on Hekman's "Truth and Method: Feminist Standpoint Theory Revisited": Truth or Justice?'; Dorothy E. Smith (1997) 'Comment on Hekman's "Truth and Method: Feminist Standpoint Theory Revisited"'.
7. This chapter is a slightly revised version of the article 'Public Sociology and Democratic Theory' (2007) published in *Sociology* 41(5), 785–98.

References

Bourdieu, Pierre (1984) *Distinction: A Social Critique of the Judgement of Taste*, trans. Richard Nice. Cambridge, MA: Harvard University Press.

Bourdieu, Pierre and Jean-Claude Passeron (1990) *Reproduction in Education, Society, and Culture*, trans. Richard Nice. London: Sage.

Bourdieu, Pierre, Jean-Claude Passeron, and Monique de Sainte-Martin (1994) *Academic Discourse: Linguistic Misunderstanding and Professorial Power*, trans. Richard Teese. Stanford, CA: Stanford University Press.

Burawoy, Michael (2005a) '2004 American Sociological Association Presidential Address: "For Public Sociology."' *The British Journal of Sociology* 56(2), 259–94.

Burawoy, Michael (2005b) 'The Critical Turn to Public Sociology.' *Critical Sociology* 31(3), 1–14.

Collins, Patricia Hill ([1990] 2000) *Black Feminist Thought: Knowledge, Consciousness and the Politics of Empowerment*. New York: Routledge.

Collins, Patricia Hill (1997) 'Comment on Hekman's "Truth and Method: Feminist Standpoint Theory Revisited": Where is the Power?' *Signs*, 22(2), 375–81. Reprinted in *Where's the Power? The Feminist Standpoint Theory Reader: Intellectual and Political Controversies*, ed. Sandra Harding. New York: Routledge, pp. 255–62.

Comte, Auguste ([1830–42] 1864) *The Positive Philosophy of Auguste Comte*, trans. Harriet Martineau. New York: Calvin Blanchard.

Dahl, Robert A. (2006) On Political Equality. New Haven, CT: Yale University Press.

Eliaeson, Sven (2006) 'Myrdal as Public Intellectual.' In: *Das Faszinosum Max Weber*, ed. Karl-Ludiwg Ay and Knut Borchardt. Konstanz, Germany: University of Konstanz Verlag.

Feyerabend, Paul (1978) *Science in a Free Society*. London: NLB, pp. 283–300.

Gramsci, Antonio (1971) 'The Intellectuals.' In: *Selections from the Prison Notebooks*, ed. Q. Hoare and G. N. Smith. New York: International Publishers, pp. 3–23.

Habermas, Jürgen ([1962] 1989) *The Structural Transformation of the Public Sphere: An Inquiry into a Category of Bourgeois Society*, trans. Thomas Burger. Cambridge, MA: MIT Press.

Harding, Sandra (1997) 'Comment on Hekman's "Truth and Method: Feminist Standpoint Theory Revisited": Whose Standpoint Needs the Regimes of Truth and Reality?' *Signs* 22(2), 382–91.

Hartsock, Nancy (1997) 'Comment on Hekman's "Truth and Method: Feminist Standpoint Theory Revisited": Truth or Justice?' *Signs* 22(2), 367–74.

Hayek, Friedrich von (1960) *The Constitution of Liberty*. Chicago: University of Chicago Press.

Hekman, Susan (1997) 'Truth and Method: Feminist Standpoint Theory Revisited.' *Signs* 22(2), 341–67.

Horowitz, David (1998) *Radical Son: A Generational Odyssey*. New York: Simon & Schuster.

Horowitz, David (2006) *The Professors: The 101 Most Dangerous Academics in America*. Washington, DC: Regnery Publishing, Inc.

Jefferson, Thomas (1779) 'Bill for Religious Freedom.' Papers 2:545. http://etext. virginia.edu/jefferson/quotations/jeff160.htm. Accessed 3 August 2006.

Merton, Robert (1942) 'A Note on Science and Democracy.' *Journal of Legal and Political Sociology* 1, 115–26. Reprinted as 'Science and Democratic Social Structure' in *Social Theory and Social Structure* ([1949, 1957] 1968), pp. 604–15.

Merton, Robert ([1949, 1957] 1968) *Social Theory and Social Structure*. New York: Free Press.

Mill, John S. (1865) 'August Comte and Positivism.' In: *The Collected Works of John Stuart Mill*. London: Trubner.

Myrdal, Gunnar ([1944] 1962) *An American Dilemma: The Negro Problem and Modern Democracy*. New York: Harper and Row.

Nilsen, Waldemar (1972) *The Big Foundations*. New York: Columbia University Press.

Pettit, Philip (1997) *Republicanism: A Theory of Freedom and Government*. Oxford: Oxford University Press.

Polanyi, Michael (1939) 'Rights and Duties of Science.' *The Manchester School of Economic and Social Studies* 10(2), 175–93.

Polanyi, Michael (1941–43) 'The Autonomy of Science.' *Memoirs and Proceedings of the Manchester Literary & Philosophical Society* 85(2), 19–38.

Polanyi, Michael (1943–45) 'Science, the Universities, and the Modern Crisis.' *Memoirs and Proceedings of the Manchester Literary & Philosophical Society* 86(6), 107–65.

Polanyi, Michael ([1946] 1964) *Science, Faith, and Society*. Chicago: University of Chicago Press.

Polanyi, Michael ([1951] 1980) *Logic of Liberty: Reflections and Rejoinders*. Chicago: University of Chicago Press.

Price, Don K. (1965) *The Scientific Estate*. Cambridge, MA: Harvard University Press.

Ranulf, Svend (1939) 'Scholarly Forerunners of Fascism.' *Ethics* 50(1), 16–24.

Rawls, John (1993) *Political Liberalism*. New York: Columbia University Press.

Schmitt, Karl ([1922] 1985) *Political Theology: Four Chapters on the Concept of Sovereignty*, trans. George Schwab. Cambridge, MA: MIT Press.

Smith, Dorothy (1993) 'Standpoint Epistemology.' *The Sociological Quarterly* 34(1), 169–82.

Smith, Dorothy E. (1997) 'Comment on Hekman's "Truth and Method: Feminist Standpoint Theory Revisited."' *Signs* 22(2), 392–8.

Stacey, Judith (2004) 'Marital Suitors Court Social Science Spin-sters: The Unwittingly Conservative Effects of Public Sociology.' *Social Problems* 51(1), 131–45.

Turner, Stephen P. and Jonathan H. Turner (1990) *The Impossible Science: An Institutional Analysis of American Sociology*. Newbury Park, CA: Sage.

Turner, Stephen (2005) 'Polanyi's Political Theory of Science.' In: *Emotion, Tradition, Reason: Essays on the Social, Economic and Political Thought of Michael Polanyi*, ed. Struan Jacobs and R. T. Allen. Aldershot, UK: Ashgate, pp. 83–97.

Turner, Stephen (2006) 'Was "Real Existing Socialism" Merely a Premature Form of Rule by Experts?' In: *Building Democracy and Civil Society East of the Elbe: Essays in Honor of Edmund Mokrzycki*, ed. Sven Eliaeson. London: Routledge, pp. 248–61.

Turner, Stephen (2007a) 'Merton's "Norms" in Political and Intellectual Context.' *Journal of Classical Sociology* 7(2), 161–78.

Turner, Stephen (2007b) 'The Social Study of Science Before Kuhn.' In: *Handbook of Science and Technology Studies*, ed. Edward Hackett et al. Cambridge, MA: MIT Press.

Weber, Max ([1917] 1949) 'The Meaning of "Ethical Neutrality" in Sociology and Economics.' In: *The Methodology of the Social Sciences*, ed. Edward A. Shils and Henry Finch. New York: Free Press, pp. 1–47.

Weber, Max ([1917] 1946) 'Science as a Vocation.' In: *From Max Weber: Essays in Sociology*, ed. H. H. Gerth and C. W. Mills. New York: Oxford University Press, pp. 129–56.

9
Varieties of Democracy in Science Policy

Erik Weber

Introduction

In *Science, Truth and Democracy,* Philip Kitcher distinguishes four forms of science policy: internal elitism, external elitism, vulgar democracy and enlightened democracy. In this chapter, I argue that Kitcher's arguments against elitism fail to show that elitism must be eliminated completely (i.e., they do not entail that all decisions about science policy should be taken democratically) and that his argument against vulgar democracy is unsound. I show that Kitcher has two characterisations of vulgar democracy, and that his argument against vulgar democracy is invalid under both characterisations. Then I argue that Kitcher's argument against vulgar democracy is in fact an argument against direct democracy and in favour of representative democracy. But that leaves open many options, especially with respect to the main topic (viz., the degree to which scientists should be involved in the democratic decision processes concerning science policy) for which Kitcher introduced the distinction between vulgar and enlightened democracy. Finally, the last section of this chapter is a constructive rather than critical one: based on my analysis of Kitcher's work, I formulate some proposals for implementing a democratic science policy.

Kitcher's views on the aims of science

In my criticism of Kitcher's proposals concerning science policy, I start from a view on the aims of science identical to his, which is why I start with a summary of these views.

Science aims at significant truths

Let us define *unrestricted intentional realism* as the view on the aims of science, which maintains that (i) the world has a mind-independent structure,

---

I thank Jeroen Van Bouwel and Rogier De Langhe for their comments on draft versions of this chapter.

and (ii) the aim of science is to describe this structure in a systematic and complete way. There is an obvious objection to unrestricted intentional realism. Philip Kitcher gave a version of this objection in *The Advancement of Science*:

> Truth is very easy to get. Careful observation and judicious reporting will enable you to expand the number of truths you believe. Once you have some truths, simple logical, mathematical, and statistical exercises will enable you to acquire lots more. ... The trouble is that most of the truths that can be acquired in these ways are boring. Nobody is interested in the minutiae of the shapes and colors in your vicinity, the temperature fluctuations in your microenvironment, the infinite number of disjunctions you can generate with your favorite true statement as one disjunct, or the probabilities of the events in the many chance setups you can contrive with objects in your vicinity. What we want is *significant* truth. (1993, p. 94)

I agree with Kitcher that objections like this one make unrestricted intentional realism untenable: we need restrictions on what is investigated and what not. The question then is: which restrictions?

In *Science, Truth and Democracy*, Philip Kitcher formulates the argument against unrestricted intentional realism as follows:

> Nobody should be beguiled by the idea that the aim of inquiry is merely to discover truth, for, as numerous philosophers have recognized, there are vast numbers of true statements it would be utterly pointless to ascertain. The sciences are surely directed at finding *significant* truths. But what exactly are these? (2001, p. 65)

Kitcher's answer to the question at the end of the quote is double. On the one hand, there is "practical significance":

> One possible answer makes significance explicitly relative – the significant truths for a person are just those the knowledge of which would increase the chance she would attain her practical goals. Or you could try to avoid relativization by focusing on truths that would be pertinent to anyone's projects – the significant truths are those the knowledge of which would increase anyone's chance of attaining practical goals. (2001, p. 65)

But for Kitcher there is more:

> Neither of these is at all plausible as a full account of scientific significance, and the deficiency isn't just a result of the fact that both are

obviously rough and preliminary. Linking significance to practical projects ignores areas of inquiry in which the results have little bearing on everyday concerns, fields like cosmology and paleontology. Moreover, even truths that do facilitate practical projects often derive significance from a different quarter. Surely the principles of thermodynamics would be worth knowing whether or not they helped us to build pumps and engines (and thereby attain further goals). Besides the notion of practical significance, captured perhaps in a preliminary way by the rough definitions given above, we need a conception of "theoretical" or "epistemic" significance that will mark out those truths the knowledge of which is intrinsically valuable. (2001, p. 65).

Kitcher is certainly not the only philosopher defending this kind of sophisticated pragmatism that leaves room for "intrinsically valuable knowledge". Carl Hempel started his famous article 'Aspects of Scientific Explanation' with the following remarks on the aims of science:

Among the many factors that have prompted and sustained inquiry in the diverse fields of empirical science, two enduring human concerns have provided the principal stimulus for man's scientific efforts.

One of them is of a practical nature. Man wants not only to survive in the world, but also improve his strategic position in it. This makes it important for him to find reliable ways of foreseeing changes in his environment and, if possible, controlling them to his advantage.

...

The second basic motive for man's scientific quest is independent of such practical concerns; it lies in his sheer intellectual curiosity, his deep and persistent desire to know and to understand himself and his world. (1965, p. 333).

Another example of this sophisticated pragmatism can be found in the epilogue of Larry Laudan's *Progress and its Problems*:

If a sound justification for most scientific activity is going to be found, it will eventually come perhaps from the recognition that man's sense of curiosity about the world and himself is every bit as compelling as his need for clothing and food. Everything we know about cultural anthropology points to the ubiquity, even among "primitive" cultures barely surviving at subsistence levels, of elaborate doctrines about how and why the universe works. The universality of this phenomenon suggests that making sense of the world and one's place in that world has deep roots within the human psyche. By recognizing that solving an intellectual problem is every bit as fundamental a requirement of life as food and drink, we can drop the dangerous pretense that science is legitimate only in so far as it contributes to our material well-being or to our store of perennial truths. (1977, p. 225)

Epistemic significance

The view presented in the previous section is prima facie attractive: it is not too narrow (science does not have to be practically useful) and not too wide (science has to relate to everyday human concerns, so it must be practically or theoretically useful). In my criticisms of Kitcher's views on science policy, I will start from a similar view on the aims of science. However, my implementation of epistemic significance will be broader than Kitcher's implementation. Kitcher uses the idea of "natural human curiosity" to clarify what epistemic significance is: "The sciences ultimately obtain their epistemic significance from the broad questions that express natural human curiosity" (2001, p. 81).

Hempel also refers to an inborn human curiosity (cf. the quote above). In my criticism of Kitcher's views, and also in the last section, I combine the idea of inborn human curiosity with other ways to justify research that is not practically significant.

Kitcher's arguments against elitism

Having summarized Kitcher's view on the aims of science, we can now turn to his proposals concerning which institutional framework for science policy is the most conducive to achieve these aims.

Four forms of science policy

In Chapter 10 of *Science, Truth and Democracy* Philip Kitcher distinguishes four forms of science policy: internal elitism, external elitism, vulgar democracy and enlightened democracy. His definitions of the first two institutional arrangements are:

> One, *internal elitism*, consists in decision-making by members of scientific subcommunities. A second, *external elitism*, involves both scientists and a privileged group of outsiders, those with funds to support the investigations and their ultimate applications (call these people "paymasters"). (2001, p. 133)

His opinion on how decisions are actually made, is this:

> I take it that the status quo in many affluent democracies is a situation of external elitism that groups of scientists constantly struggle to transform into a state of internal elitism. (2001, p. 133)

In Kitcher's view Francis Bacon (with his *New Atlantis*) was the first defender of elitism:

> The first report on science policy was written as fable. In *New Atlantis*, left incomplete after his death, Bacon offered a tale about the crew of a sailing ship, who, after various disasters, find a haven in the island of Bensalem. Here the mariners are treated with great hospitality, and they are surprised by the wisdom, generosity, and incorruptibility of the island's government. Liberality and fair-dealing are founded on the institution of an elite group of investigators, the members of Salomon's House who seek "the knowledge of causes, and the secret motions of things."[1] In the terms I've been using, they aim to achieve both epistemically and practically significant truths, and they are very clear about the character of both. (2001, p. 137)
>
> ...
>
> The work of Salomon's house is carried forward by coordinating autonomous decisions made by the members in discussion with one another. At various stages, they confer to plan the next steps to be taken (following the method for individual inquiries Bacon outlined in many places). The fellows decide what to do, what to publish, what to keep secret, and what applications to make. So we are offered an explicitly elitist vision of well-ordered science, one that takes an objectivist vision of the good that which inquiry aims. There are certain things which it is good for human beings to know because it will relieve their curiosity, certain things which it is good to know because applying the knowledge will contribute to human welfare. The wise inquirers understand these things and thus bestow all kinds of benefits on their society. (2001, pp. 137–138)

The second part of the quote reveals an important characteristic of Kitcher's argument against elitism: he assumes that elitism can only be justified by invoking an objectivist vision on what practically significant and epistemically significant truths are.

Kitcher's rejection of objectivism about values

According to Kitcher, one of the core philosophical questions with respect to science policy is "what are good ways for a community to organize inquiry if they want to promote their collective values" (2001, p. 114). This is a difficult question, because the goal which occurs in it is vague:

> When we broaden the perspective to encompass our "collective values", the goal becomes much more nebulous. How does this relate to the actual wishes and preferences of the members of a society? How are we to

integrate the preferences of different people? Can very different types of value be brought under a single measure? (2001, pp. 114–115)

A first possible way to deal with such questions is the objectivist one:

> There are two very obvious ways of approaching the problems of the last paragraph. One is to suppose that whatever preferences people actually have, whatever they think about what it would be good for them to pursue, either individually or collectively, some ends are objectively worthy and there are objective relations among these ends. Call this general perspective *objectivism about values* (*objectivism*, for short). Objectivism can concede that there are many different kinds of values, some of them practical, some epistemic, some present, some future. I may even countenance human diversity, supposing some packages of good things are better for some people, different packages for other people. But objectivists think there's a right way of trading the epistemic against the practical – and, more generally, a right way of trading various different types of values for one another – a right way of balancing the present against the future, and a right way of integrating the objective interests of different individuals. (2001, p. 115).

Kitcher formulates two objections against this objectivism:

> My doubts rest on the difficulty of divorcing what is good for a person from that person's own reflective preferences and the kindred problem of ignoring personal preferences in understanding the ways in which different distributions of goods across the stages of a person's life yield overall value. Further, I think that the general problem of understanding how to aggregate individual levels of well-being into a measure of collective welfare, in the ways objectivists propose, is extremely difficult. (2001, p. 116)

Because of these objections, he chooses a subjectivist perspective:

> Individual preferences should form the basis for our understanding of the personal good that inquiry (among other social institutions) is to promote. In moving from the individual to the measurement of value for the society, we should explicitly limit our discussions that honor certain democratic ideals. Hence my approach to the fundamental question, "What is the collective good that inquiry should promote?" will start from a subjectivist view of individual value (using personal preferences as the basis for an account of a person's welfare) and will relate the individual good to the collective good within a framework in which democratic ideals are taken for granted. (2001, p. 116)

Kitcher is convinced that Bacon's idea is utopian: elitistic decision-makers cannot know which research is good for the population for which they are responsible. He concludes from this that elitism (both the internal and the external variant) is a bad way to organise science policy.

Different types of policy questions

I think that Kitcher's arguments against elitism are generally sound, but I will argue that they fail to show that elitism must be eliminated completely: they do not entail that all decisions about science policy should be taken democratically. To see this, it is important to distinguish different types of policy questions. The most general science policy questions a government or international organisation has to answer are the following: How much money will be spent on scientific research? How much of this money should go to research with practical significance, as opposed to research that has only epistemic significance?[2] Kitcher's rebuttal of an objectivist perspective on values is a good argument to support the idea that these issues should be decided democratically, rather than in an elitist way (I will discuss this in more detail in the last section). Let us assume that it has been (democratically) decided that x per cent of the budget for scientific research should go to projects with no practical significance (only epistemic significance), while the rest goes to research that has some practical significance. Kitcher's subjectivist perspective gives an argument for dividing the latter part of the budget in a democratic way among more specific domains and among topics within domains (this will also be discussed in more detail in the last section). Here we confine ourselves to the x per cent that must be spent on research with only epistemic significance. It can be argued that an elitist procedure is the best way to distribute that money. I offer two arguments below one based on the idea of subsidiarity, one based on the idea of serendipity.

Subsidiarity

The first argument starts from natural human curiosity as implementation of epistemic significance. What are the consequences of invoking natural human curiosity as justification for doing and financing scientific research which has no practical significance? Human curiosity varies a lot across cultures and individuals. At first sight, this variety seems to constitute an argument for democratic decisions in the area of epistemically significant research. On closer inspection, this turns out not to be true. The results of the research will only satisfy the curiosity of a group of scientists who are familiar with the topic. Advanced scientific research will not satisfy the curiosity of the general public, because the general public does not understand what is written in scientific journals. Nonscientists can benefit from *practically* significant research because using the results does not require mastering the underlying theories. For instance, I can take aspirin and benefit from it without knowing anything about how it works. This kind of "detachment" is impossible with respect to curiosity. Note that I am *not* saying that the curiosity of nonscientists is not important or can be neglected: my claim is that science is not an appropriate instrument for satisfying this curiosity.

Under this assumption, the principle of subsidiarity comes into play. In a survey article on this principle, Andreas Føllesdal describes it as follows:

> The "principle of subsidiarity" regulates authority within a political order, directing that powers or tasks should rest with the lower-level sub-units of that order unless allocating them to a higher-level unit would ensure higher comparative efficiency or effectiveness in achieving them. (1998, p. 190)

The subunits can be territorial or functional (see Føllesdal 1998, p. 196). In the first case, it regulates the relation between levels in a federal state (e.g., states and the federal government in the United States; German Länder and the German Federal Republic) or an international political union (e.g., the member states of the European Union versus the European Union). In the second case, the subunits are social groups without a specific territory. For instance, in many countries, physicians and lawyers have their own professional organisation which develops a code of ethics and has institutes for enforcing it. If we assume that only scientists benefit from the result of epistemically significant research (because that research can only satisfy the curiosity of scientists) then one should allow scientists, as a subunit, to take their own decisions. This is not elitism in the sense that it entails that scientists decide what is good for other people. It is democracy combined with subsidiarity: scientists who decide about what is good for themselves, without interference of outsiders who are not stakeholders.

Before I turn to my second argument, I want to mention an assumption that I have tacitly made. I assume that the class of "research which has only epistemic relevance" is not *necessarily* empty. Kitcher says that "in many cases *though not in all*, epistemic and practical interests are interwoven" (2001, p. 76; emphasis mine). The emphasised part of the quote is crucial. If epistemic and practical interests would *always* be interwoven, my criticism would be pointless because research which has only epistemic relevance would be impossible. However, examples of past research show it is possible. Whether or not it constitutes a large part of scientific research should be a democratic decision.

Serendipity

Invoking inborn or natural curiosity is not the only way to justify doing and financing scientific research that has no obvious practical significance. Another route is to claim that there are many serendipitous discoveries. Kitcher does not consider this possibility. I consider it here because it could be a different way to argue against elitism with respect to research that has only epistemic significance.

To make this route work, it is important to distinguish between pseudoserendipity and real serendipity:

> I have coined the term *pseudoserendipity* to describe accidental discoveries of ways to achieve an end sought for, in contrast to the meaning of (true) *serendipity*, which describes accidental discoveries of things not sought for.
>
> For example, Charles Goodyear discovered the vulcanization process for rubber when he accidentally dropped a piece of rubber mixed with sulfur onto a hot stove. For many years Goodyear had been obsessed with finding a way to make rubber useful. Because it was an accident that led to the successful process so diligently sought for, I call this a pseudoserendipitous discovery. In contrast, Georges de Mestral had no intention of inventing a fastener (Velcro) when he looked to see why some burs stuck tightly to his clothing. (Roberts 1989, p. x)

Another example of pseudoserendipity is Archimedes' famous accidental discovery – by stepping into a bath – of a way to measure to volume of irregularly shaped solids (Roberts 1989, pp. 1–3). Pseudoserendipity, as defined by Roberts, means that a practically significant result which we have been looking for is obtained partially by accident. In order to justify research without clear practical significance, only real serendipity can be invoked. More specifically, we need real serendipity where the aim of the scientist is not practical at all. Roberts's book contains several examples of this kind of serendipity: besides Velcro, there is, for instance, the discovery of saccharin as a by-product of purely theoretical research (p. 151) and the discovery of X-rays by Röntgen, who had no practical applications in mind (pp. 139–143). Real serendipitous discoveries in which the scientist wanted to find something else of practical significance do not help us if we want to argue that it is worthwhile to spend money on research that has no clear practical value. For instance, William Perkin discovered the first artificial dye (mauve) while looking for a way to synthesize quinine, the malaria drug, which at that time could only be obtained from the bark of the cinchona tree (see Roberts 1989, pp. 66–70). Such cases of serendipity do not help to justify research without practical value, because Perkin aimed at finding something practically useful.

Let us now investigate the consequences for science policy of invoking real serendipity (of the right kind) as justification for doing and financing scientific research which has no practical significance. If we assume that it is impossible to judge in advance which prima facie practically useless research will in the end lead to practically significant results, then the heuristic value of research can only be established with hindsight. Considerations of heuristic value should have no place in policy decisions, because they fail to discriminate: everything can have serendipitous effects and thus turn out to have heuristic value for practically significant results. Decisions about which practically nonsignificant research should be done and financed should be made on the basis of formal criteria, not on the basis of content. With formal criteria I mean criteria which guarantee that the research results in reliable knowledge. **These criteria include** the capacities of the researchers and the methodological soundness of research proposals. Since scientists are experts in these formal criteria (for instance, they can judge the methodology proposed by other scientists), scientists should make those decisions. To put it differently, the serendipity perspective (as an argument for financing nonpractically significant research) entails no restrictions on the content of scientific research. It only entails that our science policy must lead to reliable scientific results (no fraud, no bad procedures, and so on). In order to ensure that outcome, an elitist science policy (in Kitcher's sense) is probably the best option.

Kitcher's arguments against vulgar democracy

What is vulgar democracy?

As already mentioned above, Philip Kitcher distinguishes four forms of science policy: internal elitism, external elitism, vulgar democracy and enlightened democracy. This is how he characterises the third and fourth institutional arrangement near the *end* of Chapter 10 of his book:

> A third, *vulgar democracy,* imagines that the decisions are made by a group that represents (some of) the diverse interests in the society with the advice from scientific experts. The fourth, *enlightened democracy*, supposes decisions are made by a group that receives tutoring from scientific experts and accepts inputs from all perspectives that are relatively widespread in the society: in effect, it fosters a condensed version of the process of ideal deliberation I've outlined. (2001, p. 133)

At the *beginning* of Chapter 10 of his book, Kitcher gives a different characterisation of vulgar democracy:

> There's a simple way to develop the idea that properly functioning inquiry -well-ordered science – should satisfy the preferences of the citizens in the society in which it is practiced. Projects should be pursued just in case they would be favored by a majority vote. Call this "vulgar democracy".
>
> Vulgar democracy doesn't require actual voting. Rather it offers a standard against which we can assess rival schemes for deciding which endeavors are to be undertaken. The idea of calling together the citizenry to cast ballots on each occasion of decision is evidently absurd, but vulgar democracy is only committed to seeking social arrangements (committees of

representatives, for example) that we might expect to do well at mimicking the outcomes of the expression of individual preferences. (2001, p. 117)

Kitcher rejects vulgar democracy. However, his argument against vulgar democracy is invalid under both characterisations he gives.

An invalid argument

Kitcher's argument against vulgar democracy is this:

> The most obvious deficiency, of course, lies in the fact that people's preferences are often based on impulse or ignorance and thus diverge from favoring what would actually be good for them. Only a moment's reflection is needed to see that the most likely consequence of holding inquiry to the standard of vulgar democracy would be a tyranny of the ignorant, a state in which projects with epistemic significance would often be dismissed, perceptions of short-term benefits would dominate, and resources would be likely to be channeled toward a few "hot topics." Because these consequences plainly diverge from the promotion of well-being, vulgar democracy is a bad answer to our question. (2001, p. 117)

Is this argument valid under the second characterisation of vulgar democracy? If we use that definition, the core of Kitcher's argument is this: Any decision mechanism about science policy that aims at mimicking the majority vote leads to tyranny of the ignorant.

Similar incompetence arguments have been given against universal suffrage for men (as opposed to only suffrage based on tax paid), suffrage for women, for blacks in South Africa and so on. Kitcher should have thought twice before joining this group. But more importantly: all these arguments turned out to be wrong after opponents lost the battle. Workers, women and blacks turned out to be competent voters. So under the second definition, the argument is not convincing.

Let us now look at the first definition. According to that characterisation, democratic decision-makers in a vulgar democracy receive advice from scientists. If the scientists do their job well, there is no reason to expect tyranny of the ignorant. So the argument that vulgar democracy leads to tyranny of the ignorant is not valid under this definition. A further problem with this characterisation is that the difference between vulgar and enlightened democracy becomes unclear: in vulgar democracy there is "advice" from scientists, in enlightened democracy there is "tutoring". That is the only difference in the definitions. If "tutoring" is just another word for "advice" (as some people may claim), there is no difference between vulgar and enlightened democracy. If tutoring means that the scientists educate the savage nonscientists, then enlightened democracy is elitism in disguise: only vulgar democracy is really democratic.

Direct democracy

There is a third definition of vulgar democracy (one that Kitcher does *not* give) under which the argument is valid. If we define vulgar democracy as direct democracy in all decisions, then vulgar democracy indeed may lead to tyranny of the ignorant. But that is a very trivial result. Defenders of referenda and other forms of direct democracy usually see them as correction mechanisms in a parliamentary democratic system. Kitcher's argument against vulgar democracy in science policy is in fact an argument against *complete* direct democracy in science policy (i.e., against the idea that all decisions in science policy should be taken by referenda) and in favour of representative democracy (with or without corrective referenda). But that looks like a straw man argument: it is very doubtful that anyone has ever proposed complete direct democracy in science policy.

To sum up: Kitcher's two characterisations of vulgar democracy do not result in a useful conceptual distinction. The third characterisation we considered faces the same problem. So it seems wise to give up the distinction.

Outline of a democratic science policy

How should science policy be organised? In this section, I present a brief outline. The first (and maybe most important question) in science policy is: how much of the money the state receives through taxes and other sources should go to scientific research? In a political democracy, this decision is – by definition – taken in a democratic way. It is an item in the yearly budget composed by the government and approved by the parliament. However, if we agree that the distinction between practical and epistemic significance makes sense (as mentioned at the beginning this chapter, my view on the aims of science are similar to Kitcher's, so I do think this distinction makes sense), the ideal situation is one in which the parliament explicitly decides how much money should go to research which has anticipated practical significance and research which has no anticipated practical significance. Such a decision is facilitated if the institutes that divide the money among researchers (cf. the second and third part of my proposal) are clearly distinguished. In Belgium, each university has a so-called "Bijzonder Onderzoeksfonds" (Special Research Fund) and a so-called "Industrieel Onderzoeksfonds" (Industrial Research Fund). The research financed by the latter one should result in "potentially application-oriented knowledge with an economic finality", while the first has no such restriction. The research it finances can have practical significance, epistemic significance or both. The money the funds distribute comes from the government. By allocating a certain amount of money to each type of fund, the government/parliament can influence the distribution between practically and epistemically significant research. However, the government/ parliament should have complete control over this distribution, and that is only possible if there are no funds with a "mixed" task (i.e., no funds that finance both practically relevant research and research without immediate practical relevance). In the Belgian institutional arrangement, the Industrial Research Funds decide about a subset of practically relevant research (research that leads to the production of technological artefacts and is practically relevant through this production). The Special Research Funds decide about a wide variety of research ranging from practically relevant research that is not "materializable" into technological artefacts (e.g., research on the prevention of cancer or research on global governance) to research without practical significance. This distribution of tasks is certainly not the optimal one.

The second part of my proposal is that decisions about research that is not practically significant should be taken by scientists only. My arguments for this can be found in the sections on subsidiarity and serendipity. Of course, within scientific communities, there is no reason for further elitism: scientists should democratically decide about who becomes members of boards (consisting of scientists) that decide about research projects that have only epistemic significance.

I suppose that in many countries there is room for some or a lot of democratisation in this respect. For instance, the members of the board of the Special Research Fund at my own university are elected by their colleagues, while this is not the case for the members of the scientific committees of the Research Fund – Flanders (FWO), which functions at a regional level (multiple universities) rather than at a local level (one university).

The final part of my proposal is a plea for a democratisation of the decisions about practically significant research projects. The Belgian federal government finances research that relates to its policy (through a framework programme called "Action for supporting the strategic priorities of the Federal Government"). The regional Flemish government does the same through a number of centres for policy-relevant research ("Steunpunten voor beleidsrelevant onderzoek"). This is useful, and in principle also democratically organised (the democratically elected government decides what is done). However, such systems (which I assume to exist in many other countries too) have three obvious shortcomings from a democratic point of view. First, political parties that have representatives in the parliament but are not in the government are excluded from the decision process. This is problematic, because they should have the right to require research on certain topics, for example, in order to prove that the policy of the government has failed in a certain domain. The second problem is that social movements which do not have a political counterpart (unlike, for instance, environmentalists who can call on "green" parties) are also excluded from the decision process. Third, democracy in these institutes is very indirect: the "will of the people" is represented by civil servants appointed by and reporting to the government (that there is representation is not a problem, but the chain of representations is very long). A really democratic science policy has to remove these shortcomings.

Notes

1. This phrase is taken literally from Bacon's book.
2. To avoid confusion, I want to clarify my terminology. Because – as Kitcher rightly argues – practical and epistemic significance are often interwoven, we should in principle distinguish three categories of research: research with practical significance but without epistemic significance, research with both practical and epistemic significance (Kitcher uses the human genome project as an example for this category) and research with only epistemic significance (Kitcher mentions research with superconducting supercolliders as an example in this category). When I write "research with practical significance", I mean the union of the first two categories.

References

Føllesdal, A. (1998) 'Subsidiarity.' *The Journal of Political of Philosophy* 6, 190–218.

Hempel, C. G. (1965) *Aspects of Scientific Explanation and Other Essays in the Philosophy of Science*. New York: Free Press.

Kitcher, P. (1993) *The Advancement of Science*. Oxford: Oxford University Press.

Kitcher, P. (2001) *Science, Truth and Democracy*. Oxford: Oxford University Press.

Laudan, L. (1977) *Progress and Its Problems*. London: Routledge.

Roberts, R. M. (1989) *Serendipity. Accidental Discoveries in Science*. New York: John Wiley & Sons.

10

Some Economists Rush to Rescue Science from Politics, Only to Discover in Their Haste, They Went to the Wrong Address

Philip Mirowski

Does a particular political predilection, often associated with neoliberalism, tend to foster skepticism towards modern science? In the middle of the last decade, a number of journalists thought so (Mooney, 2005); and with the election of Barack Obama, many of those same journalists rejoiced that now science would once more be restored to its rightful place in the polity. But attitudes towards both science and politics turn out to be much more entrenched, with roots running deep into theories of political economy, and thus exhibiting a more stubborn persistence than most of those commentators realized. Indeed, currently attitudes towards the politics of science are heavily bound up with the extent to which science is seen to readily respond to public 'demands', something that is frequently conflated to the extent to which science is subordinate to the dictates of the marketplace. Consequently, it will prove useful to summarize some recent developments in the economics of science to begin to comprehend why science is still under stress in the modern polity.

The new evolutionary economic epistemologists

Once upon a time, just after World War II, science was treated as a mysterious cornucopia produced by unworldly monks united by Mertonian norms and an unquenchable search for Truth. Under the influence of the atomic bomb, the computer and the space race, the general public was satisfied that scientists would keep churning out innovations that would power economic growth and the advance of civilization. This was accompanied by an older economics of science that treated the ultimate wellsprings of knowledge as a mystery, yet that nonetheless counseled state subvention under the rubric of 'public goods' to keep the fires of innovation stoked.[1]

With the oil shocks and productivity slowdowns of the 1970s, this world-view began to unravel. Various problems with the earlier neoclassical economics of science had grown so glaring and insistent that they practically conjured a sort of counter-orthodoxy from the 1970s onwards. While many of the disgruntled stewed in isolation, over the ensuing decades, the disaffected successfully gathered together a transnational epistemic community through international workshops, shared participation in (European) science policy positions and purpose-built journals. Recently, a few of the main protagonists began explicitly proclaiming the existence of a 'Stanford/ Yale/Sussex School', thereby striving to claim the mantle of the 'new orthodoxy' in the economics of science and technical innovation.[2] Some of them have been bold thinkers who have sought to integrate the history of technology (and sometimes science), and philosophical approaches to knowledge, with economic models that were often driven more by capturing the salient empirical trends than rescuing the textbook neoclassical approach. They also tended to favor a more 'organicist' and less mechanistic view of what it meant to be a science. While they may not have been intellectually all-inclusive – we shall document their antipathy to the field of science studies below – they did appeal to much wider disciplinary audiences than their counterparts, the 'regular' economists, who mostly just talked amongst themselves. Because they sought to attach themselves to more modern concepts of successful science, seeking and sometimes finding inspiration in biology, we shall dub them the 3E school (Evolutionary Economic Epistemologists), highlighting the facet of their writings most relevant to the economics of science. More to the point of our eventual verdict, they also frequently consciously set themselves in opposition to the neoliberal phalanx that sought to analyze science as a literal marketplace of ideas.[3] Yet we shall discover that merely pledging troth to the opposition is not enough; it will be necessary to thoroughly understand the intellectual sources of the neoliberal ascendancy to comprehend how comparatively easy it is to backslide into glib dependence upon the marketplace of ideas when studying science. In particular, it has become too easy to believe that the market unproblematically produces 'truth' like it produces toothpaste when one deals in economic models of science.

While it certainly helps to access their own self-characterizations when cobbling together a summary characterization of their major tenets, we shall venture a bit further to differentiate the members of the 'school' into subsets according to the extent they have forged a set of doctrines that effectively mark a departure from the neoliberal complex described in Mirowski and Plehwe (2009). In the Table 10.1, representative economists situated to the 'right' have diverged from the older orthodoxy in relatively ineffectual ways (and thus maintain a higher profile amongst economist colleagues), while those situated further to the 'left' have tried much harder to burn their bridges. Of course, the columns of the table do not sport hard-and-fast boundaries: concepts and even protagonists tend to migrate between subgroups.

Table 10.1 The 3E school

Sussex – Science Policy Research Unit [SPRU]; Scuola Sant'Anna – Pisa	Nelson and Winter Style Evolutionary Economics; National Systems of Innovation at OECD	Stanford Economic History; Santa Fe Institute
Giovanni Dosi; Chris Freeman; Paul Nightingale; Benjamin Coriat; Mariana Mazzucato; Geoff Hodgson	Richard Nelson; Sidney Winter; Franco Malerba; Stan Metcalfe	Paul David; Nathan Rosenberg; Gavin Wright; Dominique Foray; Brian Arthur; Joel Mokyr; Steven Durlauf
Technological trajectories; bounded rationality; panel data stochastic models; reject production functions	Competence theory of firm; routines as genes/memes; national systems innovation; history-friendly models; Schumpeterian creative destruction	Path dependence; QWERTY; tacit/codified distinction; optimal theory of 'open science'; Markov chains; spin glass models

It would seem reasonable to ask what intellectual commitments actually hold the various 3E contingents together as a community.[4] One of the major figures of the previous 'public good' orthodoxy, Richard Nelson, has now in fact crossed over to become one of the standard-bearers for one major battalion of 3E, indeed, more or less repudiating his previous 1959 paper (Nelson, 2006). Curiously enough, it seems that the desire to keep a channel open to the orthodox economics profession in order to preserve some semblance of theoretical legitimacy may be their broadest common denominator. This becomes apparent in the account given by Dosi, Malerba, Ramello and Silva (2006) of the 3E 'synthesis': at some level, they profess to still accept the older Arrow/Nelson portrayal of science as a 'public good' requiring public subsidy, but simultaneously invert the discredited linear model of the relationship of basic science to technology: "Practical inventions come about *before* scientific understanding of how they worked" (p. 894). They augment this with an insistence upon the 'tacit' character of some subset of knowledge, thus making it difficult for agents to render at least that portion into a conventional economic commodity. Sometimes this 'tacitness' is conflated with a purported pervasiveness of 'bounded rationality' amongst scientists, with acknowledgments to the Carnegie School of Herbert Simon. Although they profess to believe that the relationship of science to technology is so complicated that no grand generalizations can be made across the board, they also feel impelled to testify that "since the industrial revolution, the relative contribution of science to technology has been increasing" (p. 894). These seemingly contradictory propositions are said to be reconciled by means of the recognition that "technological change is an evolutionary process," although it is not at all clear if they would extend their credo to accept that science is itself also an evolutionary phenomenon.[5] What they do concede is that Science possesses a special ethos (with the obligatory genuflection to Robert Merton), which they claim consists of "(i) a sociology of the scientists' community largely relying on self-governance and peer evaluation; (ii) a shared culture of scientists emphasizing the importance of motivational factors other than economic ones; and (iii) an ethos of disclosure of research results driven by 'winner takes all' precedence rules" (p.894). Because they assert without argument that this 'Open Science' can only be carried out in a publicly [read: government] supported venue, it is a major tenet that "appropriability conditions [read: intellectual property] have at most only a limited effect on the pattern of innovation" (p. 896). Although this clashes with their renunciation of the linear model, they then insist that it is technological innovation that is the prime cause of economic growth, but it is nonmarket subsidized Open Science that is the engine of technological innovation.

Beyond the spice lent by condiments like 'bounded rationality', tacit knowledge and appeals to evolutionary theory, it may appear at the end of the day that 3E does not cash out as being really so very different from earlier conventional doctrines of the neoclassical economics of science. Of course, the further to the 'right' one negotiates the table, the more this seems the case. Sometimes that might seem to be a glaring inconsistency in their own promotional materials, and individual members of the 3E collectivity do struggle with this. However, it is also apparent they are appalled by the neoliberal turn taken by the neoclassical orthodoxy since roughly 1980, yet perhaps almost as much by the rise of the disciplines of 'science studies' with its offshoot the 'Social Construction of Technology'.[6] The 3E school undoubtedly feel their work as under threat from all sides, in the former case by people such as Hayek and Armen Alchian and Ronald Coase preaching that the marketplace of ideas is just evolution in another guise, as a prelude to the comprehensive privatization of science (Butos & McQuade, 2006); and in the latter case, from what they perceive as a pack of wild-eyed crazies treating science as just another cultural phenomenon. That is probably why they exhibit the repetitive tic of harkening back fondly to Robert Merton and Joseph Schumpeter as hallowed forebears. Merton was proponent of a version of the 'sociology of science' (now obsolete within sociology), which maintained a bulletproof screen between broader social considerations and the content of science, not to mention 'the market' and the 'invisible college'; Schumpeter was a mathematically challenged cheerleader for the Walrasian tradition coming to displace other schools of economics in the early postwar United States, while at the same time insisting that the Walrasian theory of perfect competition could not explain capitalist growth and vitality, which in his view came instead from large firms in concentrated industries organizing and funding huge programs of technological innovation, which had the consequence of setting in train massive waves of creative destruction.[7]

The 3E figures we shall encounter below have been progressively forced to straddle a sequence of intolerable contradictions, all in the name of their promotion of a 'new economics of science'. One of these key contradictions is that the logic of their position pushes them to reject Arrovian-style welfare economics, but nevertheless, they just cannot bring themselves to fully jettison the technical public good concept once and for all. It is almost as if they were bereft of a language that would allow them to adequately express the social character of science and technology. Another contradiction is the strangely symmetrical trap that everything they believe in suggests that they should pledge their allegiance to something looking very much like the linear model (reducing growth to the motor of technological change, itself reduced to breakthroughs in basic science), but one of their standard tropes is the *pro forma* rejection of the idea that science causes technological change. But the contradictions get worse. Most of the 3E cadre praise the very contingency of human history to the skies (whether it be through 'path dependency' or evolutionary non-determinism or 'bounded rationality'), but when it comes to *science* itself, they revert to approaching it as though it existed, pristine and complete, in a static Platonic world wholly detached from the social processes they describe. It appears they sometimes confuse a belief in philosophical realism with a strict quarantine of the Social from the Natural. This, in turn, has had the debilitating effect of rendering them incapable or unwilling to entertain the notion of differential economic infrastructures producing better or 'worse' quality science; for them, everything boils down to simple quantitative indices of more or less gross science for the buck. This, in turn, practically compels them to reify knowledge as a commodity in their theories, even while insisting that knowledge is context-dependent. The *sub rosa* determinism even extends to their own methodological practices, in the sense that the more successful 3E analysts have generally been avid to appropriate natural science models (spin glass, Markov diffusion, Polya urns, Fisherian natural selection, chaotic dynamics) and import them wholesale into economics to explain firm growth and decay, the 'trajectories' of technological development, and more. The reader might query: why doesn't social science possess its own Platonic truths? The 3E tradition still tends to treat 'knowledge' held by others in a crudely instrumentalist manner, revealing their roots in conventional economics; but when their own work is threatened to be framed in a symmetric fashion (Albert & Laberge, 2007), they suddenly morph into the noblest of disembodied intellects, interested in only 'pure truth'. Perhaps the most grating contradiction (and the most salient for our present concerns) is their insistence that 'appropriability conditions' generally do not determine levels and rates of innovation (and thus the neoliberal marketplace of ideas is denounced as irrelevant), but then are stalwart in their political project to pronounce that there is nonetheless something fundamentally unhealthy about the modern commercialization of science. Clearly many 3E members have trouble imagining what 'harm to science' actually looks like. The only disability they seem to imagine is not purchasing sufficient amounts of science, though they have no way of identifying when enough is enough. They suspect that some parties may be politically harmful to science, but can't seem to reconcile this with their economic theories of democracy. While there are many fascinating aspects of the 3E movement that could legitimately be covered from our current perspective, in this chapter, we will focus upon a few key chinks in the armor of 3E theorists in setting themselves up as self-appointed defenders of the virtue of science from the depredations of the neoliberal detractors of science and the modern privateers of knowledge.

Nelson and Winter 'refuting' neoliberal evolution

We might begin by wondering: where did the doctrine of 'evolution' in 3E come from? It was not, as one might initially suspect, from familiarity with Darwinian biology, or from previous Veblenian Institutionalist economics. The origin of modern appeals to evolution in economics is admitted by a large cohort of 3E theorists[8] to be Armen Alchian's paper 'Uncertainty, Evolution and Economic Theory', which appeared in the 1950 *Journal of Political Economy*. Alchian (later a member of the Mont Pèlerin Society and a major neoliberal theorist) framed his analysis as constituting a continuation of basic neoclassical theory as found in Alfred Marshall: "we shall be reverting to a Marshallian type of analysis combined with the essentials of Darwinian evolutionary natural selection" (1950, p. 213). His paper begins with a critique of profit maximization as a general explanatory principle under uncertainty: "where foresight is uncertain, 'profit maximization' is meaningless as a guide to specifiable action" (1950, p. 219).[9] Although the paper claims its motives as essentially conservative, namely, to bring economics back to 'a Marshallian type of analysis', what is noteworthy is the relative dearth of any citations to Darwin, or indeed, to Marshall. 'Biology' is discussed only in the most cursory superficial terms, absent even the bare bones of the by-then conventional litany of variation, inheritance and specification of selection mechanisms, except in a final passage in the conclusion: "The economic counterparts of genetic heredity, mutations and natural selection are imitation, innovation and positive profits" (1950, p. 220). However, what is being imitated is left wholly without specification, later sowing doubt as to whether it exhibited sufficient invariance and persistence required for evolution; variation is likewise left unexplained. Even the selection mechanism(s) are left unspecified, with everything nominally collapsed to a single undefined profit index.

The basic message was this: Economic theory will still be able to validate neoclassical outcomes if the market mechanism is understood to be functioning as a selection mechanism, selecting over different manifestations of firm (or individual) behavior with profit as the motive underlying the evolutionary process. This is asserted to lead to the conclusion that economic success is not based upon motivation, but rather upon results; individual behavior without "rationality, motivation and foresight" was said by Alchian to be perfectly consistent with aggregate neoclassical predictions (p. 221). How this was to be squared with what one found actually written in Marshall was left unexplored.[10] Whether evolution would actually attain a maximum was never demonstrated; it was simply presumed. Uncertainty precludes individual intentional maximization, but is claimed to be consistent with different forms of learning through imitation of successful rules or individual trial and error learning.

Starting with Sidney Winter, many of the 3E proponents set out on their careers motivated by a burning desire to refute the proposition that 'evolution' would validate the standard neoclassical model, as well as the neoliberal doctrine of an efficient marketplace of ideas. They regard themselves as bringing a more psychologically plausible version of limited rationality to the fore, and forestalling Panglossian prescriptions that this is the best of all possible worlds. But their own relationship to the natural sciences in this quest has been ambivalent, to say the least. And yet, because the terms of the argument had initially been set by neoliberals like Alchian and Milton Friedman, the neoliberal version of 'thin evolution' can be observed becoming ingrained in the writings of many 3E scholars. The confusion starts with the notion of 'selection', a term that too readily conjures up connotations of one-shot selection of elements from a given set, resembling the 'selection' of commodities in a marketplace (Knudsen, 2002). That is emphatically *not* the meaning it betokens in biological evolutionary theory, where it denotes a winnowing down of differential replication of descent through modification. Alchian himself omits a number of important components of Darwinian evolution, including: (a) specification of the conditions of relative stability of the firms vis-à-vis their environment; (b) restrictions upon the complete plasticity of the firm in its reactions to the environment, through his confusion of genetic transfer with 'imitation'; (c) provision of any guarantee of the maintained identity of either the replicator or the interactor (by eliding the distinction expressed in biology by genotype vs. phenotype); and (d) an account of the emergence and operation of the selection rules. Because of this sloppiness, Alchian seems not to appreciate that 'natural selection' *per se* nowhere constitutes the sum total of evolutionary dynamics, but serves as only one component (others include drift, neutral mutation, frequency dependence of co-evolution, etc.). In 'thin evolution', both descent and differential replication are treated as mere secondary considerations, and the phenotype/genotype distinction is suppressed, as a prelude to neglecting or ignoring them completely, producing an entirely 'ahistorical' travesty of evolution. It is no accident this ends up resembling the extremum models in physics so favored by the neoclassical orthodoxy. Thus, in 'thin evolution', there abides no real evolutionary change, as understood by sophisticated proponents of Darwinism.[11]

This observation begins to bite not only Alchian, but also many members of 3E, especially whenever they feel moved to reconcile their exuberant evolutionary rhetoric with other scholars' understandings of biological transformation and historical contingency. The repeated protestations of Richard Nelson and Sidney Winter that they need not be held faithful to any biological analogies in their work is one very pertinent instance of their rather sketchy treatment of Alchian's four issues mentioned above.[12] In their simulation work (1982), they still resorted to neoclassical production functions to model firm technologies; and they never adequately explained why it was that the 'routines' that they claimed were the very essence and identity of firms were invariant with respect to changes in firm personnel, or indeed, more stable than the rate of change of the environment. If commercial knowledge sported an ineffable and tacit character redolent of Michael Polanyi's retort to J. D. Bernal, that too was no accident, because Winter has admitted he had been put onto Polanyi's writings by Thomas Kuhn (Augier, 2005, p. 349). In effect, Winter was portraying corporate management in much the same way that Polanyi portrayed his scientists, further cementing the underlying bidirectional analogy to a marketplace vis-à-vis science. Consequently, all the prognostications that firms would not generally be observed to maximize were hardly better founded than Alchian's insistence that they *would* maximize something. Thin evolution rules nothing out, and nothing in. The transubstantiation through which all these trademark neoliberal doctrines (from Alchian, Friedman, Polanyi) would get reprocessed into something that would refute a neoliberal approach to knowledge was the avowed goal of Nelson and Winter, a goal that seems little closer today than it was back in Winter (1964), with the finish line further obscured by misleading appeals to evolution. It had consequences for the way this has played out in the economics of science in Nelson (2006), where he fails to derive his current proscription that the privatization of university science has gone "too far" from his trademark theory of technological advance as an evolutionary process: "I want to stress that my position that the results of university research generally should be open and available for all to use should not be confused with an argument that universities should be 'ivory towers'" (2006, p. 915). Other than nostalgia for the Cold War, it's not at all clear what 3E suggests the modern university could or should aspire to be, according to Nelson. After all, thin evolution embodies no inherent teleology, and Nelson has always used market models of one sort or another to understand how knowledge is related to scientific research.

Paul David's wobbly dependence upon path dependence

The full complement of the quandaries of 3E can be found in the work of Stanford economic historian Paul David. David likes to characterize his life's work as follows: "Most of my working life has been spent as a member of a small dissenting sect – advocating *historical* economics – within the more-or-less tolerant (albeit largely indifferent) society of academic economists" (1993a, p. 210). The reason for their tolerance lies in the meaning of 'historical' in David's writings. Far from being the full frontal challenge that Nelson and Winter hoped to mount, David has always presented it as "something additional, and for many, something new has to be learned. That 'something' can stand alongside neoclassical economic analysis, and so enhance one's appreciation of the special features distinguishing that paradigm from what may be called *historical economics*" (2001, p. 16). 'History' is here intended to augment the orthodox model of knowledge and not derange it in any fashion. No wonder David's writings have turned out to have been the most extensively cited of the entire 3E cohort. 'Historical economics' thus boils down to a very limited number of ideas in David's scheme of things: first off, it designates 'path dependency', but most people have heard of it in its popular incarnation as QWERTY; second, it seems bound up with a central distinction between tacit and codified knowledge; and finally, when it comes to the economics of science, it means a rational-choice account of the origins and success of certain institutions underpinning scientific research. How then can this limited repertoire be reconciled with the broader concerns of the 3E cadre, much less serve to guarantee the centrality of 'history' to economic discourse? Through the magic elixir of – what else? – evolution! "The concept of path dependence refers to a property of contingent, non-reversible dynamical processes, including a wide array of biological and social processes that can properly be described as evolutionary" (2001, p. 15). Let us take a peek at how the marriage of evolution and harmony is wrought.

It should be stated at the outset that Paul David conceives of 'historical economics' as primarily 'theoretical', waving away the hint of oxymoron with the confidence of someone who appreciates the limits to what economists will put up with. He has conveniently provided both a negative and a positive definition of his trademark concept of path-dependency (2001, p.19):

Negative definition: Processes that are non-ergodic, and thus unable to shake free of their history, are said to yield path dependent outcomes.

Positive definition: A path dependent stochastic process is one whose asymptotic distribution evolves as a consequence (function of) the process's own history.

Over his career, David has wavered between endorsing the notion that his abstract model of path dependence is best exemplified by a pure stochastic process, based upon a Markov chain, which may not even exhibit fixed absorbing states (due to time-varying parameters), and the more conventional (from a neoclassical vantage point) notion of a deterministic system that exhibits multiple equilibria, with the equilibrium arrived at accounted for by sensitive dependence upon initial conditions (viz., the erstwhile 'chaos' theory once fashionable at the Santa Fe Institute).[13] David tended to fudge the very real distinctions between the two options in the 1990s by insisting he would find path dependence in situations where there was some special predilection of a technology for 'increasing returns', or the idea that the more a technology was used, by its very nature, the more productive it would become. This would involve making use of neoclassical production functions – characteristically, David has never rejected this fundamental analytical device.[14] (When he began finding path dependence in ideas and institutions, he has since played down this aspect of the argument.) Either option would serve to underwrite his assertion that "the core concept of path dependence as a dynamic property refers to the idea of history as an irreversible branching process" (p. 23), but his real predicament comes when David attempts to meld his enthusiasm for path dependence with his adherence to neoclassical microeconomics. It is symptomatic of David's version of the economics of knowledge that his every article contains at least one endorsement of the public good argument as found in his Stanford colleague's paper (Arrow 1962). That means, of course, that on some level he still believes in static neoclassical welfare economics, as well he might, given he has the unfortunate habit of finding 'market failures' everywhere he casts his glance.[15] So is 'historical economics' just a synonym for 'pervasive market failures'?

'Welfare' or 'market failure' in Walrasian general equilibrium can only be defined relative to an equilibrium the system is destined to occupy, and that's where all the trouble resides. Because there is no canonical dynamics in the Walrasian tradition, it is a tricky matter to state definitively whether a particular notional dynamical system is or is not conformable to the neoclassical tradition. Worse, because in the Walrasian model uniqueness of the equilibrium cannot be demonstrated (outside of some narrow special cases), there is no sense that a particular identifiable equilibrium is the terminus of any given adjustment process. Nonetheless, neoclassical price theory was inspired by deterministic physics of conservative systems, and the Arrow tradition has been associated with trying to come up with a deterministic dynamics of adjustment, so while David's latter option (chaos and multiple equilibria) enjoys some tenuous purchase in postwar neoclassical price theory, his option of nonergodic stochastic processes has none. The entire neoclassical approach is predicated upon the commandment that the economic system exhibits substantial determinacy, which permits scientific statements to be made concerning its operation. This means it must exhibit 'reversible' choices, because otherwise, the whole neoclassical notion of 'substitution' would lose all meaning. Most card-carrying neoclassical economists would therefore reject out of hand the 'anything can happen' approach of exfoliating irreversible branching indeterminacy; and this accounts for David's own indeterminacy in his two interpretations of the meaning of 'path-dependence' in the economy. Obfuscation in reference makes for tolerance in economics departments.

I know the reader must look askance on the last two paragraphs as excessively and numbingly abstract, having little to do with the issues that occupy this volume; but luckily we have the antidote, because they have been played out in a very concrete example of 'path-dependence' for which David has become famous. As one of his 3E critics has pointed out, "the tricky methodological issue [is] 'how do you recognize path-dependency when you see it?' After all, we just observe *one* simple path, since we can only see the one history that has occurred" (Dosi, 1997, p. 1543).[16] In a study published in 1985, David announced he had found a tangible example of historical path dependence in technology adoption staring anyone in the face who found themselves shackled to a keyboard in order to express themselves – that is, nearly every academic.

The QWERTY keyboard layout, which is what I am typing on now, was devised and created in the 1860s by the creator of the first modern typewriter, Christopher Sholes, a newspaper editor. Initially, the characters on the typewriters he invented were arranged alphabetically, set on the end of a metal bar that struck the paper when its key was pressed. However, once an operator had learned to type at speed, the bars attached to letters that lay close together on the keyboard became entangled with one another, forcing the typist to manually unstick the type bars. David reveled in the irony that Scholes then created the QWERTY layout in order to slow the typist down and to prevent nearby letters from jamming the platen mechanism. Remington typewriter designs subsequently became the standard for Anglophone cultures, and consequently, people learned to touch-type by memorizing the QWERTY layout. It was important for David to insist that this was a concertedly inferior technology, in the first instance for ergonomic reasons, but eventually, since the keyboard was maintained long after the type-bar technology had been displaced by IBM Selectrics, and then, of course, the personal computer, because it by then was a total mismatch to improvements in text inscription technology.

To seal his case, David then described a history of alternative typewriter layouts that were 'better'. For instance, QWERTY inconveniences are minimized by any of the numerous competing keyboard designs that concentrate the most common English letters onto the home row. To motivate our comparison, consider the two alternative keyboard layouts in Figure 10.1.

Qwerty

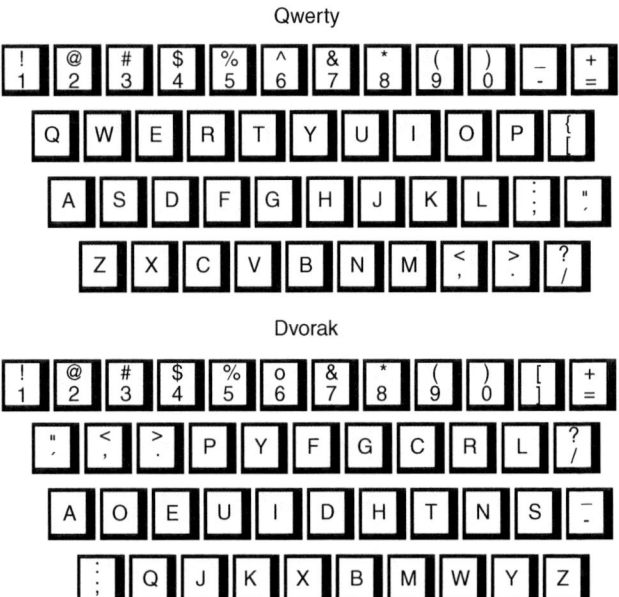

Dvorak

Figure 10.1   Rival keyboard layouts

The candidate that David highlighted, the Dvorak keyboard, introduced in the 1930s, devotes the home row to 9 of the 12 most common English letters – including all five vowels and the three most common consonants (T, H, N) – while the 6 rarest letters (V, K, J, X, Q and Z) are relegated to the bottom row. As a result, 70 percent of typing strokes remain on the home row, only 22 percent are on the upper row, and a mere 8 percent are on the hated bottom row; thousands of words can be typed with the home row alone; reaches are five times less frequent than in QWERTY typing, and big row jumps hardly ever happen. Another easily understood drawback of the QWERTY keyboard has to do with alternation of hands. Whenever the left and right hands type alternate letters, one hand can be getting into position for the next letter while the other hand is typing the previous one. You can thereby fall into a steady rhythm and type quickly. Yet QWERTY typing (in English) tends to degenerate into long one-handed strings, especially strings for the weak left hand. The Dvorak keyboard instead forces you to alternate hands frequently. It does so by placing all vowels plus Y in the left hand, but the 13 most common consonants in the right. As a result, not a single word or even a single syllable can be typed with the right hand alone, and only a few words can be typed with the left hand alone. Because alternation would be preferable from David's perspective, the persistence of QWERTY looks like 'market failure'.

The point here is not to delve too deeply into what actually constitutes an efficient keyboard,[17] but instead to use his own exemplar to illustrate the way that David tends to portray 'path dependence'. It is critical to observe that he treats 'efficiency' here as though it were a relatively fixed benchmark independent of the social processes that produce and sustain the technology of the typewriter, as well as the organizational framework within which it operates. Ergonomic efficiency persists intact throughout history in David's world, so that when type-bar jamming eventually goes the way of buggy whips, then keyboard efficiency reverts to its 'true' equilibrium, but the actual state of affairs remains 'locked in' on an inferior equilibrium. One recognizes this now as the static 'multiple equilibria' version of path dependence. That of course would not be the case if he had really adhered more faithfully to his alternative dictum of time-dependent nonergodic stochastic processes: one could easily imagine that both ergonomics and touch-typing practices could have co-evolved over time, partly in reaction to the shift from type bar to Selectric to laptop, and partly due to changes in relative prices, but perhaps also due to other keypad experiences in different contexts (think of bank ATMs, or touch-screen keypads), resulting in some third yet unimagined keyboard that would reconcile any number of considerations into an easy and error-free typing experience. In this scenario, there would be no such thing as a 'true' keyboard (because perhaps that's what evolution really looks like). But in that case, what would count as 'efficient' would display no fixity, constantly changing at the same rate as the rate of change of the environment, with the consequence that there could never be any such thing as a flat-out 'inferior' technology. Absent a clean Platonic metric of 'inferiority', a ghostly archetype perduring throughout history, there would of course be no definitive 'market failure'.

The QWERTY story gained such traction in the early 1990s that some other economists sought to produce their own Old Testament versions of 'historical economics' conforming more closely to neoclassical theory. The arguments of Stan Liebowitz and Steven Margolis (1990; 1995; 2000) were not willfully 'evolutionary', but they did serve to reveal David's Achilles Heel, in that they homed in on this presumption of a fixed given independent metric of 'inferiority'. Conveniently for our frame tale, their work was funded and supported by well-known neoliberal think tanks and foundations.[18]

Liebowitz and Margolis (1990) aimed much of their counterargument at the alleged superiority of the Dvorak keyboard, arguing the documents David cited for the superiority of the Dvorak keyboard were based on dubious or flawed experiments. Other experiments that Liebowitz and Margolis cited supported the conclusion that it could never be profitable to retrain typists from QWERTY to the Dvorak keyboard. Moreover, Liebowitz and Margolis found ergonomic studies that concluded the Dvorak keyboard offers at most only a 2 to 6 percent efficiency advantage over QWERTY, which they suggest was not large enough to overcome costs of conversion.

Although Liebowitz and Margolis never addressed David's claims about the role of lock-in through third-party typing instruction, they did argue that firms had sufficient opportunities to offer training in conjunction with sales of typewriters, so that non-QWERTY keyboards would not have been seriously disadvantaged. This left open the question of how much typing instruction was initially offered directly by suppliers, as Liebowitz and Margolis suggest could have happened, and how much was offered by third parties using QWERTY, as David argued did happen. In summary, Liebowitz and Margolis did uncover historical evidence that early typewriter manufacturers competed vigorously in the features of their machines. They inferred, therefore, that the reason that typewriter suppliers increasingly supported and promoted QWERTY must have been that it offered an overall competitive advantage as the most effective system available according to market logic. We were restored to living in the best of all possible worlds, and it elicited a sigh of relief.

There were a myriad of reasons the QWERTY controversy attracted the extraordinary amount of attention it did throughout the 1990s. One was that the parable became bound up with the anti-trust case against Microsoft, as a way to indict that firm as promoting lock-in to their own inferior technology, although this time around it was an operating system rather than just a keyboard. Another was that it became conflated with the 'increasing returns' fad in Paul Romer's 'new growth theory' (Warsh, 2006). And then there was a *frisson* of enthusiasm at the Santa Fe Institute that 'increasing returns' would somehow midwife an entirely new way of doing economics. Most of these conditioning factors were transitory, however. The deeper and more persistent itch that kept the QWERTY controversy festering was the following question: what precisely is the best way to account for historical technological change while remaining faithful to the neoclassical model? The emotions roiling just beneath the surface were best exposed in Paul David's screed "at last, a remedy for chronic QWERTY-skepticism!" (1999).[19]

David correctly apprehended that his critics had taken him at his word: to talk in terms of public goods one must have at one's disposal some concept of welfare losses, and that means that one could only talk about 'market failures' in quantitative terms of more or less welfare loss, at least if you wanted to hold a genteel conversation with the neoclassical profession. Everything else, as the critics sneered, was just urban myths and bogus anecdotes. But then, what did all that palaver about 'path dependency' really amount to, other than effective repudiation of the pivotal concept of neoclassical equilibrium? His response was most curious. David began by tossing off the observation that everyone knows that Pareto efficiency is just an idealization and historically inaccessible. (This presumes a level of irony I have not frequently encountered in orthodox economists.) Then, if one allowed there were multiple equilibria, which one would be the 'right' benchmark in a dynamical sense? The Olympian analyst might take a position on this, but the agents themselves would only intermittently be aware of what was at stake in moving from one such benchmark to another. Finally, he conjures 'increasing returns' as the *deus ex machina* to banish any obligation to identify benchmarks under any circumstances: "where the source of non-ergodicity in the system happens also to be one of the large class of (convexity-destroying) conditions that undermine attainment of Pareto efficiency in a competitive market regime, market failure becomes a likely outcome." Note well the illegitimate conflation of the two divergent interpretations of path-dependency earlier mooted: one where there are lots of equilibria, and the other where there are none. The bottom line is that Paul David absolves himself and his followers from having to justify any statements concerning the metric of inferiority: "the question of what is and is not 'inefficient' is quite separable from the question of the usefulness of the conceptual frame work of path dependence." Well, not to rain on the 3E parade, but that's just not plausible, if the whole purpose of the exercise was to *augment* the neoclassical organon with a little dash of history. Reprising Dosi's question: how do we know 'path-dependence' if we see it? What if the concept is a will-o-the wisp? Facile appeals to more empirical work can't solve this problem.[20] Maybe history can't be worn so lightly as just another flashy accessory to intellectual dominance.

In the remainder of this section, I will suggest that Paul David (and most of his cadre at Stanford) in fact don't do justice to history; and that is because they approach history like economists, not historians and science studies scholars. A quick way to sketch the difference is to quote David Bloor, a famous defender of the so-called 'Strong Programme' of science studies, when he characterizes the 1970s dustup between *philosophers of science* and historians of science, recounting:

Bullying attacks from philosophers who wanted to reify and ring-fence 'reason' and who effectively treated the 'internal logic' of science as if it were an ahistorical, self-propelling and autonomous force. For these philosophers, society came into the story merely as the precondition of science, not as a constituent of knowledge. Society facilitated or impeded the autonomous growth of rationality. Fortunately, the historians did not allow themselves to be browbeaten. ... For the historians, then, the deployment of reason by the scientists posed a problem, and the answer to this problem was neither obvious nor provided by the scientists themselves. The problem was: why do the proffered reasons typically convince some scientists but not other scientists? ... The historians' response was to contextualize the ideas and arguments under examination. They looked for the intellectual traditions into which the competing parties fell, the institutions with which they were associated, and the goals and interests that might be behind the argument. In this way they could illuminate the unspoken assumptions and the taken-for-granted tendencies that were at work. (Bloor, 2007, p. 221)

If we look at the QWERTY case, we observe David treating 'path dependence' and lock-in just like the philosophers were treating 'reason': things are bestowed with the power to be self-propelling and autonomous. That is not the way that historians of science and technology tend to argue; but it is a characteristic vice of the 3E approach to the economics of science. Some other 3E economists further to the left of Table 10.1 have begun to comment on this (Nightingale, 2008). Once the critics unpacked path dependence, they discovered that the more it was contextualized, the more it would lose its potency as Implacable Historical Imperative. If the trajectory of a technology is not something that 'society' or 'the economy' simply facilitates or frustrates, but turns out to be a function of the way in which 'technical superiority' is itself constructed by the victors, then the lifeline to neoclassical economics is irretrievably lost.

As David says, this is just not a local quibble about a relatively insignificant fable of the keys. We should take Paul David at his word: the real crux of the matter is how to make use of history to discuss technological change, and by implication, the operation of science. As he says, "Path dependence, at least to my way of thinking, is therefore about much more than the process of technological change, or institutional evolution, or hysteresis effects and unit roots in macroeconomic growth. The concepts associated with this term have implications for epistemology, for the sociology of science, and cognitive science as well" (2001, p. 33). Throughout the 1990s David claimed to be forging a "new economics of *science*" along the same lines as his earlier work on technological change; yet, if anything, the quest to treat 'knowledge' the way he had reified QWERTY only exacerbated the incongruity of producing abstract ahistorical accounts of the dynamics of historical change.

Around the time that the Cold War 'public good' justification of science policy was meeting its donnybrook, Paul David began casting about for ways to 'revise' his story in order to take into account neoliberal accusations of the 'inefficiency' of state-subsidized academic science and answer the charge that flaws in the system were causing the United States to falter in 'competitiveness'. Characteristically, he joined forces with a rather conventional orthodox welfare economist, Partha Dasgupta, to produce a more up-to-date model – meaning, essentially, one that now had to resort to Nash equilibria in noncooperative game theory, which had displaced Walrasian general equilibrium as the state of the art in neoclassical microtheory in the 1980s – to concede some ground to the neoliberals, but nevertheless, to 'defend' state-organized science. In a paper published in 1994, David and Dasgupta allowed that there were "numerous inefficiencies" in the current system, but warned that, "in reforms proposed to promote knowledge transfers between university-based open science and commercial R&D...there are no economic forces that operate automatically to maintain dynamic efficiency in the interactions of these two spheres".[21] Although this paper has become widely cited as providing the basis of a 'new economics of science' as well as a banner for 3E, two things immediately stand out: one, none of those authors actually uses the formal model or even subjects it to any serious scrutiny; and two, in practice, the effect of the paper has been to extend a benediction to all manner of endorsements of the commercialization of science, contrary to the protestations actually found in the paper. It is worthwhile to spend a little time examining it, if only to fend off the objection that "economists no longer believe in simple stories of the marketplace of ideas."

The paper starts out by lamenting "that elaboration of economic analyses of technology was for some time allowed to run far ahead of the economics of science" (p. 223). This was interpreted as the analysts somehow *neglecting* to treat science as a purely economic process, when in fact, it was their *intent* all along to relegate science outside the profane sphere of pecuniary pursuit. David and Dasgupta proposed to rectify the situation by appropriating a model from the economics of industrial organization and applying it to scientific research, in effect asserting there was no ontological difference when it came to the marketplace of ideas. (So much for the brave insistence on separate but equal spheres.) The ideas that they imported from industrial organization were: (1) that there was a 'principal-agent' problem in science, namely, the patron or employer had trouble 'monitoring' the productivity of the employee-researcher; (2) that there were 'economies of scale' in knowledge production [thus permitting David to insert his trademark theme of increasing returns – but not, significantly, path dependence]; and (3), it was an economic calculation to decide what, if anything, to disclose to the public and what to keep secret. The next step was a little odd. They define information as "knowledge reduced and converted into messages that can be easily communicated" (p. 226), and that this reduction process is the means whereby one turns knowledge into a commodity. They then equate codified knowledge/information with the prior Arrow definition of a 'public good', and contrast it with 'tacit knowledge', which remains outside the commercial sphere, citing Michael Polanyi for support. Now, one might think that codified knowledge precisely would *not* qualify as a public good given the massive fortification of intellectual property rights going on in that era (reducing intellectual artifacts to a digital format was the preferred prelude to engaging in a little primitive accumulation in the 1990s), but the authors waved this objection away in a footnote. Then, in an absolute inversion of what Polanyi actually wrote, they associate codified knowledge with the products of an academic sphere of 'Science', and tacit secret science with the alternate parallel sphere of 'Technology' (but really: commercialized science). You have to have both, they intone (although never say why); but "the boundary line between tacit and codified knowledge is not simply a matter of epistemology; it is a matter also of economics, for it is determined endogenously by the costs and benefits of secrecy in relation to those of codification" (p. 233). Once again we note the trademark David habit of positing a static welfare optimum for all time, and then introducing a smattering of 'history' to justify divergences from that ahistorical optimum. Just in case the reader has missed the moral of the exercise in the thickets of the model: in this 'new economics of science', there exists an *optimal level of the commercialization of science*, which is to be determined by a virtual marketplace (those costs and benefits), themselves governed by neoclassical welfare criteria of Pareto optimality, except where those pesky public goods gum up the works.[22] The Market giveth all good things, and the Market dictates where the Market should rule: *this is pure pasteurized neoliberalism*. How this was supposed to 'defend' open science from the encroachments of commercialization turned out to be more a matter of legerdemain centered upon arbitrary distinctions imposed upon the presumed 'thing-like' knowledge, than it is anything having to do with what had been actually going on in the advanced economies since the 1980s.

What about the game theoretic model? One can speculate this was added more for ceremonial legitimacy than analytical explication, given the sloppiness with which issues of equilibrium were deployed. Basically, they posited a prisoner's dilemma game played by scientists, where the suboptimal Nash equilibrium involved all researchers keeping their findings secret (=tacit), while the 'cooperative' solution of providing codified publication would only be enforced by Mertonian 'norms' of priority rule 'credit', which were supposed the sole province of academic science. (The empirical fact that corporate scientists regularly publish in academic outlets would apparently refute the model, but seems not to have occurred to anyone who has read this paper then or since.) Cooperative behavior is also said to be reinforced by the threat of being treated as an 'outcast'. Without coming right out and saying so, the authors continually insinuate that self-interest can make academic science work just fine; which has the supplementary benefit that neoclassical models of self-interest are therefore appropriate when attempting to model science as a social phenomenon – with everything incongruous that implies about considerations of methodological individualism as adequate to describe science, the lack of any palpable real price system for much 'knowledge', ascriptions of godlike infinite rationality and powers of calculation beyond the Turing Machine, and every other criticism that would have been launched against this option prior to 1980. Indeed, the authors admit at the end of their paper that, far from clarifying any of the challenges facing science in the 1990s, their paper was best approached as yet another instance of crass economic imperialism:

> [T]here are numerous features of the reward system and characteristic institutional structures of open science in the modern West that give rise to resource misallocations and static inefficiencies in the conduct of basic and applied research. Correspondingly, there may be a wide field here for economists specializing in contract theory and institutional mechanism

design to familiarize themselves sufficiently well with the detailed inner workings of Science. ... Although the institutions and social norms governing the conduct of open science cannot be expected to yield an optimal allocation of research efforts, they are functionally quite well suited to the goal of maximizing the long-run growth of the stock of scientific knowledge. ... Those same institutions and social norms, however, are most ill suited to securing a maximal flow of economic rents from the existing stock of scientific knowledge by commercially exploiting its potential for technological implementations. (pp. 240–241).

Perhaps I might translate for those unfamiliar with such periphrasis: Economists don't know much about Science, but no matter, since we propose to treat it like a market, no one will punish us for doing so. Sure, academic science doesn't make money, but it sure is good at churning out published papers. We will tell you that churning out published papers is a good thing for corporate technological innovation, but we can't tell you precisely how that works. But not to worry: in the past, the market did a good job of deciding what proportion of 'public' published papers to commercialized science was 'optimal'; and once you hire enough of us economists as consultants, we will get back to you on what that proportion should look like nowadays. And by the way, don't expect us to have anything to say about the quality of the knowledge that comes from rampant commercialization of science; like Merton, we only deal in abstract social structures, as though they were entirely decoupled from what counts as knowledge. But whyever should you worry about this? The Invisible Hand guarantees that the contest of self-interested scientists inevitably leads to superior quality knowledge for all.

It is probably apparent that I think this is a monumentally cynical paper: cynical in its message, cynical in its timing, cynical in its effect on the economics profession. But my personal impressions are neither here nor there: the significance of the paper inheres in what it facilitated, and what it condoned. While pleading the opposite, it succeeded in according legitimacy to the neoliberal approach to the marketplace of ideas for a whole generation, well beyond the circle of original neoliberal economists. It has managed to produce its anticipated effect: since it was published, similar exercises have mushroomed in the economics literature. Paper after paper applying rational choice market models to science without apology can now be found sprinkled throughout specialized outlets like the *Journal of Technology Transfer* (published by the Association of University Technology Transfer Managers) to respected orthodox economics journals like the *American Economic Review, Journal of Economic Literature* and *Journal of Economic Theory*.[23] Economists with their little two-person games and static optimization exercises now blithely pronounce on what should happen to 'Science' in the twenty-first century in the United States and the world without shame, pity, competence or any sense of reflexivity.

It might, however, be useful to follow up on the impact of Stanford-style economics of science upon the rest of the 3E movement. The initial reaction to David and Dasgupta (1994) was to quibble with their definitions of 'tacit' vs. 'codified' knowledge.[24] It was quite extraordinary, but perhaps not altogether seemly to observe a gaggle of economists disputing fine points of epistemology in economics journals. Beyond the coarse presumption on the part of the David camp that 'knowledge' can be conveniently reified as a thing ("codified knowledge is similar to a commodity...there is a general trend towards the increasing codifiability of knowledge" [Foray, 2004, pp. 74, 86]), the subsequent controversy served to explicate how codification of knowledge fit snugly into David's grand scheme of path dependence in history:

> [C]odified knowledge can be a potent 'carrier of history'—encapsulating influences of essentially transient and possibly extraneous natures that were present in the circumstances prevailing when the particular codes took shape. Having that power, it can become a source of 'lock in' to obsolete conceptual schemes, and to technological and organizational systems that are built around those. (Cowan, David & Foray, 2000, p. 248)

Nevertheless, it became apparent rather early on that David had been collapsing too many different attributes of knowledge within his single tacit/codified continuum in order to make the original model work. For instance, the issue of ease (or cost) of expression was not at all isomorphic to the ability to keep something secret as opposed to engage in disclosure. Likewise, there was no reason that 'tacit' had to be associated with industrial science – surely academic research exhibited as many, if not more, tacit competences (which was, after all, Polanyi's original point; not to mention the stock in trade of such science studies scholars as Harry Collins). Also, there was no isomorphism between tacit/codified and private/collective knowledge, once one took into account changing legal treatments of intellectual property over time.

Paul Nightingale (2003), in one of the most perceptive critiques, pointed out that the antonym of 'tacit' is 'conscious', not 'codified'; David had been confusing distinctions more appropriate for the computer with those better suited to human cognitive abilities. Once the tyro economic epistemologists took this critique on board, then it was clarified that David and Dasgupta had built their whole model around treating tacit vs. codified knowledge as substitutes (and hence subject to their neoliberal trade-off), whereas, in fact, it would be much more plausible to regard them as complements. In inscribing this current argument in bits right now, I am not just 'objectifying' it, but I am also using the experience to discover what it is I actually really think – that is, my own thought processes are, in part, opaque to me, and not just others. Nightingale further suggests that those such as David and Foray get the neuropsychological evidence backwards, arguing that, "as we learn, knowledge that was unarticulated becomes codified in problem-solving heuristics and routines. The neurological evidence shows that learning can start with written rules, but as we learn, the rules become redundant and we rely on neurological processes that cannot be articulated...the assumption that the extent to which tacit knowledge gets codified, in its non-trivial sense, is determined by marginal costs commits a category mistake that confuses knowledge, which is a *capacity*, with information, which is a *state*" (2003, p. 172). Here we witness the difference between economists inventing arbitrary abstract categories willy-nilly just to get their model to 'work', and scientists (Nightingale trained as a chemist) taking advantage of access to other sciences to clarify and strengthen the sources of validity of their own theoretical tradition.[25] It is no accident that Nightingale hails from SPRU-Sussex.

The European Wing at SPRU and Sant'Anna

I would not want to end this chapter on the bleak note that nearly every-thing economists have said about science ends up supporting the neoliberal privatization of science. If there is a bright patch on the horizon, it appears in the 'left' wing of Europeans within 3E, if only because they have the least invested in protecting neoclassical economics from corruption from other schools of thought. A number of figures, either trained at SPRU Sussex or are affiliated with it, have produced a raft of insightful studies on the economics of science, and in particular with regard to the current regime of globalized commercialized research, although it would be exaggeration to claim that they have as yet produced a theoretical synthesis of commensurate ambition to that of Paul David or Sid Winter. It is noteworthy that they tend to be affiliated with the 'business schools' that have sprung up across the European terrain in the last few decades, or else find themselves housed in public policy centers. Their remit is the explicit understanding of science and technology in the contemporary political economy, and not necessarily in keeping the contemporary economics discipline happy. Perhaps because of the relative novelty of their disciplinary identities, they are inclined to produce fine-grained empirical studies and small-t theories, and therefore do not tend to garner the attention they may deserve on this side of the Atlantic. Here we shall just briefly touch on the work of three representatives: Giovanni Dosi, Benjamin Coriat and Paul Nightingale.[26]

Giovanni Dosi is currently Professor of Economics at the Sant'Anna School of Advanced Studies in Pisa, where he also leads the Laboratory of Economics and Management (LEM). He is also the impresario behind the attempt at synthetic packaging of the 'Stanford-Yale-Sussex' school of 3E quoted at the outset of this chapter, and so manifestly he is a proponent of a Big Tent approach to building epistemic communities, in contrast to the categories and distinctions we have sought to draw in this chapter. In that same spirit, Dosi has been relatively promiscuous in the past in his appeals to 'evolutionary economics' as something to which all manner of disgruntled scholars could subscribe (Dosi & Nelson, 1994). He started out back in the 1980s seeking to document the existence of 'technological trajectories' possessing a quantum of inertia, motivated by comparison to Thomas Kuhn's 'paradigms' (1962). This may sound a lot like David's 'path dependence' of technologies, and indeed, Dosi also pioneered the application of Polya urn experiments to models of technology before David (Dosi & Kaniovski, 1994). However, of late, he has grown much more skeptical of David's amalgam of neoclassical economics and 'history': "One must sadly admit that evolutionary arguments have been too often used as *ex post* rationalizations of whatever the observed phenomena ... the relevance of path dependent selection of relatively 'bad' institutional setups and technologies remains a highly controversial question" (Dosi & Castaldi, 2004, pp. 21, 23). Increasingly, he has also taken aim at production functions as flawed tools for analyzing technological change and the role of knowledge. As he craftily suggests, the correct answer to the question "how do I bake a cake?" is most emphatically not to respond "maximize price times output minus price times inputs" (Dosi & Grazzi, 2006, p. 179). In Coriat and Dosi (1998), he has argued that dependence upon 'routines' and tacit knowledge as fundamental theoretical entities in Nelson and Winter had the unfortunate effect of rendering working rules in exaggerated individualist and cognitive formats, and that the role of power relations in knowledge had been neglected (to avoid perturbing the neoclassical economists).

More recently, Dosi has been moving much closer to the science studies position, although he has not been very explicit about this. For instance, while many in the 3E camp treat institutions as parameterizing state variables, derived from the rational choice of profit-seeking agents, formed by explicit 'constitutional' conventions, he argues that institutions shape the cognitive and behavioral identities of agents, with institutional structures prior to the definition of agent self-interest; also, he insists that much institutional structure derives from unintended consequences of processes that resemble notions of 'self-organization' and emergence found, for instance, at the Santa Fe Institute. Rather than treat firms as a nexus of rational contracts, subject to the frictions of 'transactions costs' as has become commonplace in the neoclassical tradition, he stresses firms as exhibiting a structural persistence beyond rational organization, often constituting what sorts of economic exchange are possible or feasible. Most important for this volume, he has taken to insisting that power relations cannot be simply reduced to 'asymmetric information' or other adventitious asymmetries, but constitute an essential feature of organizations, and as such cannot be portrayed merely as artifacts of economic exchange. You can't rescue science from politics with an economic model that exiles power to an unexplained residual.

The direct relevance of Dosi for the economics of science consists of two recent projects: (1) the quest to argue that recent initiatives by the European Union to imitate the neoliberal reorganization of research characteristic of the United States will have deleterious consequences for the future of research in Europe (Dosi, Marengo & Pasquali, 2006; Dosi, Llerena & Sylos Labini, 2006); and (2) a trademark project of the SPRU-Sant'Anna axis to rethink the track record of one of the key sectors of the modern regime of science organization, the pharmaceutical/biotechnology sector (Dosi & Mazzucato, 2006).

Another important figure in the left 3E community is Benjamin Coriat, professor at University of Paris-13. Coriat's connections to the French 'regulation' school has predisposed him to be much more willing to think in terms of 'regimes' of science organization and governance than many of the other 3E figures, breaking out of the endemic narrow blinkered perspective of so many analysts. He fostered new ways to apply this organizing principle to the modern regime of commercialized science; his work, along with the Parisian historian of science Dominique Pestre, has been inspirational for Mirowski (forthcoming a). Coriat combines close attention to specifics of the changes in intellectual property rights (Coriat & Orsi, 2002; Coriat, 2002) with a comparative perspective on the ways in which the biotechnology sector resembles and differs from other nominally 'hi-tech" sectors (Orsi & Coriat, 2005; Coriat, Orsi & Weinstein, 2003). He also was the first to point out that there were severe problems showing up in the biotechnology sector, counteracting the reflex tendency to treat it as a wonderful cornucopia and the wave of the future of scientific research, a stance that has now spread to semiconventional wisdom within the Harvard Business School (Pisano, 2006).

The third and perhaps most important figure of the left 3E is Paul Nightingale, Senior Lecturer at SPRU Sussex. Nightingale was trained as a chemist, but later became interested in the political economy of science (Nightingale & Martin, 2004). Nightingale has revealed both theoretical inclinations, which involve integration of real cognitive science and the social studies of science literature into theories of the organization of scientific research (1998; 2003; 2004), and deft empirical capacity, which has been trained with great effect upon the pharmaceutical industry (2000; Nightingale and Martin, 2004; Hopkins et al., 2007). His role in sounding the tocsin over unfounded enthusiasm regarding the 'biotechnology revolution' can provide a salutary prophylactic for universities and governments in their unseemly haste to jump on the 'the new knowledge economy' bandwagon.

At a deeper level, Nightingale (2004) revisits the Cold War argument that the inherent unpredictability of science must necessarily lead to a version of the 'public goods' story in the economics of science. Instead, he argues for the "moderate-unpredictability thesis – that when dealing with economically important technologies, predictability is typically, but not always, constructed" (2004, p. 1264). Here the knowledge attained does not exist independent of the social means used to arrive at it. Hence technology can exist as a relatively autonomous body of knowledge distinct from science, "because it is possible to know how to produce effects without knowing how those effects are produced" (pp. 1271–2). However, in order to deploy a technological capacity to produce and maintain a constructive predictability, a whole range of distinctively scientific capacities and abilities are required. This can be used to explain why firms cannot simply apply science 'off the shelf' to their problems, and why internal R&D capacity is a complement, not an alternative, to external market sources. Technological capabilities are costly to set up, and have long gestation periods, are generally localized and mostly non-traded. They look a lot like what is frequently dubbed 'the science base': an interlocking set of institutions that meld research, education, development, politics, publication and recruitment. Nothing could be further from the neoliberal notion of a 'marketplace of ideas'. As Nightingale puts it, "The justification for the public funding of science is not based on unquantifiable, abstract theory or market failure arguments about the provision of public goods. It instead revolves around the empirical requirement for the infrastructure needed to produce technology and allow markets to work" (2004, p. 1278).

One reason Nightingale is so important for the future of the politics of science is that his insights can be applied to the economics profession and its relationship to science, a matter he has given some thought to of late (2008). One lesson of this chapter might be that economists have repeatedly approached their contemplation of science more as a technological phenomenon than in a scientific spirit. As Nightingale says, they have been more concerned with producing a preconceived effect (say, a certain policy) than coming to understand how a theory of science, one which would venture into all manner of unforeseen and ill-defined areas, might be built up from multiple acts of empirical scrutiny. For example, as we have stated, Nelson (1959) and Arrow (1962) were more concerned with justifying a big increase in federal (military) subsidy of what they called 'basic' science in the Cold War than they were with understanding how corporations, governments and universities might fit together in a flourishing and stable American science base. It was easier for them to simply *presume* science was just another instance of a market phenomenon, full stop. Their job at that juncture was to satisfy their patrons at RAND. Later on, David and Dasgupta (1994) produced a model to assuage our fears concerning the ongoing privatization of science; we could simply choose how much privatized science from column A and public science from column B that we wanted in any given university, in something like a meta-market. Likewise, there was no serious attempt to explore how technology (e.g., the Internet) and the economy (e.g., de-industrialization) were feeding back on the very structure and content of globalized commercial scientific research.

This raises the possibility that there is an unhealthy codependency between the science policy community and the economics of science that needs to be weakened before we can really come to understand our current predicaments. Langdon Winner has raised this possibility in congressional testimony:

> Indeed, there is a tendency for career-conscious social scientists and humanists to become a little too cozy with researchers in science and engineering, telling them exactly what they want to hear (or what scholars think the scientists want to hear). Evidence of this trait appears in what are often trivial exercises in which potentially momentous social upheavals are greeted with arcane, highly scholastic rationalizations. How many theorists of intellectual property can dance on the head of a pin? (Winner, 2003)

The message of Nightingale (2004) is that a better understanding of the relationship of science to technology today would need to acknowledge that some technological innovation can certainly happen in the absence of a developed science base, and that science, technology and the economy do not interact in any single fixed lockstep fashion. However, certain kinds of scientific infrastructure and institutional ties to other power centers (corporations, governments, universities, NGOs) are required to render the world sufficiently predictable so that innovation can itself proceed in a somewhat predictable manner. Without a vibrant science base, no one would recognize those random acts of genius for what they were.

Thus, when journalists rejoice that the new Obama administration has restored science to its former exalted status, they miss all the ways in which 30 years of neoliberal science policy both in Europe and America have irreversibly altered the science base. Lifting a few restrictions on stem cell research does nothing to address the privatization of the universities, the extreme fortification of intellectual property, the extensive corporate outsourcing of R&D to China, the electronic fencing of a century of journal archives, the disappearance of the tenure track career path, and last but not least, the confusion of reliable knowledge with what is readily sold in the marketplace.

Notes

1. This is described in greater detail in Mirowski (forthcoming a).
2. See Dosi, Malerba, Ramello and Silva (2006); Dosi, Llerena and Sylos-Labini (2006); Verspagen and Werker (2003); Sharif (2006); Amin and Cohendet (2004), chap. 2; and Albert and Laberge (2007).

3. See, for instance, Francois Chesnais, former head of DSTI at the OECD: "We were fighting neo-liberalism... We were doing this in spite of Margaret Thatcher and Ronald Reagan, so we were saying 'national' when the trend was already saying governments must bow out" (quoted in Sharif, 2006, p. 753). Winter (1963) began as a critique of Friedman on evolutionary analogies; his work on tacit knowledge was heavily informed by Michael Polanyi.

4. As a bit of supporting evidence, Verspagen and Werker (2003) report that in their survey, the three top academic units in their specialty reported by their respondents are SPRU, MERIT at Maastricht University and Scuola Sant'Anna. I have found it difficult to associate any special intellectual position with MERIT; and it is clear from other evidence that Stanford plays a major role in both the concepts and legitimation of 3E. Their survey reports the top three journals in the estimation of their respondents were *Research Policy, Industrial and Corporate Change* and *Journal of Evolutionary Economics*. The intellectual issues are clarified in Nightingale (2008).

5. The hesitation of 3E to cite a philosopher like Hull (1988) in pursuit of their evolutionary credentials may be due to the fact his version of evolution is much closer to the neoliberal account of Hayek than they can feel comfortable with. It does seem 3E is much more inclined to treat a 'technology' as an entity having a discrete identity over time than they would concede for a 'scientific theory'.

6. The canonical citations on social studies of technology are Pinch and Bijker (1987), MacKenzie (1996) and Pickering (2005). Science studies began as a rebellion against the Mertonian version of the sociology of science. Examples of 3E economists disparaging science studies include Aghion et al. (2006); Cowan, David and Foray (2000); and Shi (2001, p. 12).

7. Generally, appeals to Schumpeter do not automatically win 3E adherents credibility and respect from the orthodoxy. See, for instance, Solow (2007): "He [Schumpeter] makes the rather paradoxical argument that, with long habit, even the process of innovation becomes routinized and depersonalized, and therefore weakened. Can he really have believed that successful capitalism is *essentially* a romantic virtue?" It does, however, nicely illustrate their own love/hate relationship with neoclassical economics.

8. The testimony of Sidney Winter as to this inspiration can be found in Augier (2005). Further consideration of events and ideas surrounding Alchian, Friedman and Hayek's forays into 'evolution' can be found in Mirowski **(forthcoming b)**.

9. Alchian echoed this point a few years later at RAND (where he was employed) in critique of the validity of system analysis (which had been the font of Arrow-style economics of technology): Because of uncertainty, "the consequences of a specified action cannot be uniquely identified in advance, [and] there is not available any generally accepted rule for rational behavior. Therefore, "it is difficult to tell whether a successful person – one who has a good record of making good decisions – owes his success to blind chance, or to excellent forecasting ability, or to excellent analytic ability which enables him to maximize for specified forecasts. ... if our difficulty lies in poor forecasting ability, the impact of system analysis is reduced" (Alchian & Kessel, 1954, p. 5). See also Alchian (1953): "We cannot estimate what the most efficient kind of weapon will be. ... We cannot sensibly choose to do research in order to later be able to develop and produce the 'optimum' weapon system. We simply do not have the required degree of foresight nor the ability to determine what we shall be able to learn and know" (p. 5–6).

10. An example of non-Marshallianism; Marshall (1920, p. 5) clearly thought that individual motivation and foresight made an important difference to what he

regarded as an evolutionary process, and identified "self-reliance, independence, deliberate choice and forethought", rather than competition, as the "fundamental characteristics of modern industrial life". In Alchian's disdain for the historical text, we also discover a nascent Chicago practice.

11. As long ago as 1984, I insisted that there was no serious theory of evolution in Nelson and Winter (see Mirowski, 1988, pp. 166–7). That case was subsequently made with much greater specificity for both Alchian and Nelson and Winter more recently by Knudsen (2002). The fact such critiques have been repeatedly ignored within 3E suggests that it is not strictly the logic of evolutionary explanation that constitutes the major motivation for their appeals to evolution.

12. For specific pleas, see Nelson and Winter (1982) and Nelson (2007). In their original book: "We emphatically disavow any intention to pursue biological analogies for their own sake, or even for the sake of progress toward and abstract, higher-level evolutionary theory that would incorporate a range of existing theories. We are pleased to exploit any idea from biology that seems helpful in the understanding of economic problems, but we are equally prepared to pass over anything that seems awkward, or to modify accepted biological theories in the interest of getting better *economic* theory" (1982, p. 11). My favorite example of deep confusion regarding evolutionary selection is Sidney Winter (1987). There, as per standard 3E practice, he admits there is no legitimate genetic inheritance or sexual recombination in his version of evolutionary economics: this is our definition of 'thin evolution'. But then he proceeds to assert that economics need not depend upon biology for its evolutionary concepts. "In the most abstract terms, an evolutionary process is a process of information storage with selective retention" (p. 614). Then using a misleading example of a library, he goes on to assert that the crux of evolution has to do with the definition of an equivalence class where elements are close copies of one another. Winter manages to get all three components of evolution wrong, as already noticed by Vromen (1995). By trying to deny importance of genome, he confuses equivalence with invariance, and conflates genotype/phenotype distinction. By stressing similarities, he represses importance of mechanisms of maintenance of variance. Selection then becomes any distinction made by some external agent, becoming confused with "how the entity is viewed by the analyst". These contortions just can't get the components down right: "inheritance, fitness and selection".

13. David actually attempts to disengage path dependence from deterministic chaos (2001, p. 21), but there were numerous instances in his writings where he made the connection directly (1993b, p. 30): "In case of deterministic systems the property of path-dependence manifests itself most immediately through the outcome's extreme sensitivity to initial conditions."

14. Early in his career (1975, p. 2), he suggests one should use them 'circumspectly'. More recently (David & Wright, 2003), he was still conducting quantitative exercises with 'total factor productivity' measures.

15. "It can be argued that [David] never departed from mainstream economics other than in commitment to the consequences of market imperfections in the form in which he has recognized them. Much [of his] writing in the historical economics vein is about adding history to the otherwise unquestioned apparatus of mainstream economics" (Fine, forthcoming). In this, he resembles his erstwhile Stanford colleague Joseph Stiglitz (2008).

16. See also Gavin Wright (1997, p. 1564): "more typically we cannot say whether a path dependent outcome is better or worse than some historical alternative, because the alternative would have entailed a journey down a learning curve

that was never undertaken and is therefore not known ... it may be impossible in the end to show that the result is inefficient."

17. Or even appear to endorse the concept. Anyone trying type on a Blackberry or iPod or other hand-held device must know what I mean.

18. See the documentation at http://som.utdallas.edu/capri/index.htm, last accessed July 2007. Liebowitz and Margolis later drew an analogy between their attack on QWERTY and Ronald Coase's attack on the public good concept by means of the history of lighthouses (David, 1999). This was more relevant for the track record of the economics of science than perhaps David was willing to admit.

19. Unusually for David, this appears never to have been published, but can be retrieved at http://eh.net/Clio/Publications/remedy.shtml (last accessed July 2007). All quotes in the next paragraph are from this document.

20. David has begun to acknowledge this in a backhanded fashion: "Whether a dynamic process in a given economic setting is or is not path-dependent remains an empirical question, and one that is frequently not so simple to resolve" (2007).

21. Although the original article was published in 1994, all quotes in the next few paragraphs are from the reprint in Mirowski and Sent (2002); this quote was from p. 220. Documentation of all the hand-wringing over 'competitiveness' in the United States in the 1980s is provided by the article by Sheila Slaughter and Gary Rhoades in the same volume.

22. I want to remind the reader once again of something David wrote when he wasn't trying quite so hard to ingratiate himself with the economic orthodoxy: "the Pareto efficiency locus – and its dynamical counterpart – remains a counter-factual idealization, at best" (1999, p. 4). This of course implies that no one would ever be able to find that optimal point; rather, this was an exercise in abstract justification of the existence of commercialized science, and an attempt to save the public good story.

23. See, for instance, Stephan (1996), Brock and Durlauf (1999), David and Keely (2003), and Kahn et al. (1996).

24. See, for instance, Cowan et al. (2000), Ancori et al. (2000), Nightingale (2003) and Foray (2004). David apparently had become worried in the interim he had left a gaping opening through which all manner of opponents to neoclassical economics could tumble through by merely admitting the existence of tacit knowledge; see Cowan et al. (2000, p. 218).

25. This difference itself is a symptom of the extent to which orthodox economists have been imbued with the neoliberal tradition over the postwar era. The instrumentalist approach to models dates back to Milton Friedman's famous methodology essay of 1953. It is far less common in the natural sciences.

26. Others deserving honorable mention are David Tyfield, Jane Calvert, John Abraham, **Mariana Mazzucato**.

References

Aghion, Philippe, Paul David and Dominique Foray (2006) 'Linking Policy Research and Practice in STIG Systems.' Paper presented to *SPRU 40th anniversary* conference, University of Sussex.

Albert, Mathieu and Suzanne Laberge (2007) 'The Legitimation and Dissemination Processes of the Innovation System Approach' *Science, Technology and Human Values* (32), 221–49.

Alchian, Armen (1950) 'Uncertainty, Evolution, and Economic Theory.' *The Journal of Political Economy* (58), 211–21.

Alchian, Armen (1953) 'Systems Analysis – Friend or Foe?' RAND D-1778.

Alchian, Armen and Reuben Kessell (1954) 'A Proper Role for Systems Analysis.' RAND D-2057.

Amin, Ash and Patrick Cohendet (2004) *Architectures of Knowledge: Firms, Capabilities and Communities.* Oxford: Oxford University Press.

Ancori, Bernard, Antoine Bureth and Patrick Cohendet (2000) 'The Economics of Knowledge: The Debate about Codification and Tacit Knowledge.' *Industrial and Corporate Change* (9), 255–87.

Arrow, Kenneth (1962) 'Economic Welfare and the Allocation of Resources for Invention.' In: *The Rate and Direction of Inventive Activity*, ed. Richard Nelson. Princeton, NJ: Princeton University Press. Reprinted in Mirowski & Sent (2002).

Augier, Mie (2005) 'Why is Management an Evolutionary Science? An interview with Sidney G. Winter.' *Journal of Management Inquiry* (14), 344–54.

Bloor, David (2007) 'Ideals and Monisms: Recent Criticisms of the Strong Programme in the Sociology of Knowledge.' *Studies in the History and Philosophy of Science* (38), 210–34.

Brock, William and Steven Durlauf (1999) 'A Formal Model of Theory Choice in Science.' *Economic Theory* (14), 113–30.

Butos, William and Thomas McQuade (2006) 'Government and Science: Dangerous Liaison?' *Independent Review* (11), 177–208.

Coriat, Benjamin (2002) 'The New Global Intellectual Property Rights Regime and Its Imperial Dimension.' Paper presented to *BNDS Seminar*, Rio de Janeiro.

Coriat, Benjamin and Giovanni Dosi (1998) 'Institutional Embeddedness of Economic Change.' In: *Institutions and Economic Change*, ed. K. Nielsen and B. Johnson. Cheltenham, UK: Edward Elgar.

Coriat, Benjamin and Fabienne Orsi (2002) 'Establishing a New Intellectual Property Rights Regime in the United States.' *Research Policy* (31), 1491–1507.

Coriat, Benjamin, Fabienne Orsi and Olivier Weinstein (2003) 'Does Biotech Reflect a New Science-based Innovation Regime?' *Industry and Innovation* (10), 231–53.

Cowan, Robin, Paul David and Dominique Foray (2000) 'The Explicit Economics of Knowledge Codification and Tacitness.' *Industrial and Corporate Change* (9), 211–53.

David, Paul (1975) *Technical Change, Innovation and Economic Growth.* New York: Cambridge University Press.

David, Paul (1985) 'Clio and the Economics of QWERTY.' *American Economic Review* (75), 332–7.

David, Paul (1993a) 'Path-Dependence and Predictability in Dynamic Systems with Local Network Externalities: A Paradigm for Historical Economics.' In: *Technology and the Wealth of Nations*, ed. D. Foray and C. Freeman. London: Pinter, pp. 208–31

David, paul. (1993b) "Historical Economics in the Long Run." In: *Historical Analysis in Economics*. Ed. Grahame Snooks. London: Routledge.

David, Paul (1999) 'At Last, a Remedy for Chronic QWERTY-Skepticism!' Unpub. paper presented to European summer school in Industrial Dynamics.

David, Paul (2001) 'Path Dependence, Its Critics and the Quest for Historical Economics.' In: *Evolution and Path Dependence in Economic Ideas*, ed P. Garrouste and S. Ioannides. Cheltenham, UK: Elgar, pp. 15–40.

David, Paul (2007) 'Path Dependence–A Foundational Concept.' *Cliometrica* (1), 91–114.

David, Paul A. and Louise C. Keely (2003) 'The Economics of Scientific Research Coalitions.' In: *Science and Innovation: Rethinking the Rationales for Funding and Governance*, ed. Aldo Geuna, Ammon J. Salter and W. Edward Steinmueller. Cheltenham, UK: Elgar.

David, Paul and Gavin Wright (2003) 'General Purpose Technologies and Surges in Productivity.' In: *The Economic Future in Historical Perspective*, ed. Paul David and Mark Thomas. Stanford, CA: Stanford University Press.

David, Paul and Partha Dasgupta (1994) 'Toward a New Economics of Science.' *Research Policy* (23), 487–521.

Dosi, Giovanni (1997) "Opportunities, Incentives and the Collective Patterns of Technological Change.' *Economic Journal* (107), 1530–47.

Dosi, Giovanni and Carolina Castaldi (2004) 'The Grip of History and the Scope for Novelty.' LEM working paper.

Dosi, Giovanni and Marco Grazzi (2006) 'Technologies as Problem-solving Procedures and Technologies as Input-Output Relations.' *Industrial and Corporate Change* (15), 173–202.

Dosi, Giovanni and Y. Kaniovski (1994) 'On Badly Behaved Dynamics: Some Applications of Generalized Urn Schemes to Technological Change.' *Journal of Evolutionary Economics* (4), 93–123.

Dosi, Giovanni, Llerena, Patrick & Sylos-Labini. (2006). 'The Relationships Between Science, Technologies and Their Industrial Exploitation: An illustration through the Myths and Realities of the So-Called "European Paradox."' *Research Policy* (35), 1450–64.

Dosi, Giovanni, Franco Malerba, Giovanni Ramello, and Francesco Silva (2006) 'Information, Appropriability and the Generation of Innovative Knowledge Four Decades after Arrow and Nelson.' *Industrial and Corporate Change* (15), 891–901.

Dosi, Giovanni, Luigi Marengo, and Pasquali Corrado (2006) 'How Much Should Society Fuel the Greed of Innovators?' *Research Policy* (35), 1110–21.

Dosi, Giovanni and Mariana Mazzucato, eds. (2006) *Knowledge Accumulation and Industry Evolution: The Case of Pharma-Biotech*. Cambridge: Cambridge University Press.

Dosi, Giovanni and Richard Nelson (1994) 'An Introduction to Evolutionary Theories in Economics.' *Journal of Evolutionary Economics* (4), 153–72.

Fine, Ben (Forthcoming) "David and the Two Goliaths."

Foray, Dominique (2004) *The Economics of Knowledge*. Cambridge, MA MIT Press.

Friedman, Milton (1953) *Essays in Positive Economics*. Chicago: University of Chicago Press.

Hopkins, Michael, Paul Martin, Paul Nightingale, Alison Kraft and Surya Mahdi (2007) 'The Myth of the Biotech Revolution: An Assessment of Technological, Clinical and Organizational Change.' *Research Policy* (36), 566–89.

Hull, David (1988) *Science as a Process*. Chicago: University of Chicago Press.

Kahn, James, Steven Landsberg and Alan Stockman (1996) 'The Positive Economics of Methodology.' *Journal of Economic Theory* (68), 64–76.

Knudsen, Thorbjorn (2002) 'Economic Selection Theory.' *Journal of Evolutionary Economics* (12), 443–70.

Kuhn, Thomas (1962) *The Structure of Scientific Revolutions*. Chicago: University of Chicago Press.

Liebowitz, Stanley J. and Stephen E. Margolis (1990) 'The Fable of the Keys.' *Journal of Law and Economics* 33, 1–25.

Liebowitz, Stanley J. and Stephen E. Margolis (1995) 'Path Dependence, Lock-In, and History.' *Journal of Law, Economics, and Organization* 11, 204–26.

Liebowitz, Stanley J. and Stephen E. Margolis (2000) *Winners, Losers, and Microsoft*. Oakland, CA: The Independent Institute.

MacKenzie, Donald (1996) *Knowing Machines*. Cambridge, MA: MIT Press.

Marshall, Alfred (1920) *Principles of Economics*. London: Macmillan.

Mirowski, Philip (1988) *Against Mechanism*. Totawa, NJ: Rowman & Littlefield.

Mirowski, Philip (forthcoming a) *ScienceMart™: The New Economics of Science*. Cambridge, MA: Harvard University Press.

Mirowski, philip (forthcoming, b) "On the Orgins (at Chicago) of Some Species of Neoliberal Evolutionary Economics" in: Rob van Horn, ed. *Building Chicago Economics*. New York: Cambridge University Press.

Mirowski, Philip and Dieter Plehwe (2009) *The Road from Mont Pèlerin: The Making of the Neoliberal Thought Collective*. Cambridge, MA: Harvard University Press.

Mirowski, Philip and Esther-Mirjam Sent, eds. (2002) *Science Bought and Sold*. Chicago: University of Chicago Press.

Mooney, Chris (2005) *The Republican War on Science*. New York: Basic.

Nelson, Richard (1959) 'The Simple Economics of Basic Research.' *Journal of Political Economy* (67), 297–306. Reprinted in Mirowski & Sent (2002).

Nelson, Richard (2006) 'Reflections on "The Simple Economics of Basic Scientific Research": Looking Back and Looking Forward.' *Industrial and Corporate Change* (15), 903–17.

Nelson, Richard (2007) 'Universal Darwinism and Evolutionary Social Science.' *Biology and Philosophy* (22), 73–94.

Nelson, Richard and Sidney Winter (1982) *An Evolutionary Theory of Economic Change*. Cambridge, MA: Harvard University Press.

Nelson, Richard (2006) 'Reflections on "The Simple Economics of Basic Scientific Research": Looking Back and Looking Forward.' *Industrial and Corporate Change* (15), 903–17.

Nightingale, Paul (1998) 'A Cognitive Theory of Innovation.' *Research Policy* (27), 689–709.

Nightingale, Paul (2000) "Economics of Scale in Pharmaceutical Experimentation," *Industrial and Corporate Change*, (9): 315–59.

Nightingale, Paul (2003) 'If Nelson and Winter Were Half Right about Tacit Knowledge, Which Half?' *Industrial and Corporate Change* (12), 149–83.

Nightingale, Paul (2004) 'Technological Capacities, Invisible Infrastructure and the Un-social Construction of Predictability.' *Research Policy* (33), 1259–84.

Nightingale, Paul (2008) 'Meta-Paradigm Change and the Theory of the Firm.' *Industrial and Corporate Change* (17), 533–83.

Nightingale, Paul and Paul Martin (2004) 'The Myth of the Biotech Revolution.' *Trends in Biotechnology* (22), 564–9.

Orsi, Fabienne and Benjamin Coriat (2005) 'Are Strong Patents Beneficial to Innovative Activities: Lessons from the Genetic Testing for Breast Cancer.' *Industrial and Corporate Change* (14), 1205–21.

Pickering, Andrew (2005) 'Decentering Sociology: Synthetic Dyes and Social Theory.' *Perspectives on Science* (13), 352–405.

Pinch, Trevor and Wiebe Bijker (1987) 'The Social Construction of Facts and Artifacts.' In: *The Social Construction of Technological Systems*, ed. W. Bijker, T. Hughes and T. Pinch. Cambridge, MA: MIT Press.

Pisano, Gary (2006) 'Can Science Be a Business?' *Harvard Business Review* 84(10), 114–25.

Sharif, Naubahar (2006) 'Emergence and Development of the National Innovation Systems Concept.' *Research Policy* (35), 745–66.

Shi, Yanfei (2001) *The Economics of Scientific Knowledge*. Cheltenham, UK: Elgar.

Solow, Robert (2007) 'Heavy Thinker.' *New Republic*, July.

Stephan, Paula (1996) 'The Economics of Science.' *Journal of Economic Literature* (34), 1199–1235.

Stiglitz, Joseph (2008) 'The End of Neo-Liberalism.' *New Europe*, July.

Verspagen, Bart and Claudia Werker (2003) 'The Invisible College of the Economics of Innovation and Technical Change.' Eindhoven Center working paper 03.21.

Vromen, Jack (1995) *Evolution in Economics*. London: Routledge.

Warsh, David (2006) *Knowledge and the Wealth of Nations*. New York: Norton.

Winner, Langdon (2003) Testimony to the Committee on Science of the US House of Representatives, 9 April.

Winter, Sidney (1963) 'Economics, Natural Selection, and the Theory of the Firm.' *Yale Economic Essays* (4), 225–74.

Winter, Sidney (1987) 'Natural Selection and Evolution.' In: *The New Palgrave*. London: Macmillan.

Wright, Gavin (1997) 'Toward a More Historical Approach to Technological Change.' *Economic Journal* (107), 1560–6.

Part V

(Future) Obstacles to the Social Sciences and Democracy

11
Fuller and Mirowski on the Commercialization of Scientific Knowledge
Francis Remedios

Introduction

As a problem for science studies, the commercialization of scientific knowledge is characterized as whether scientific knowledge is a public good, like health care and education, or a positional good, a good whose value allows for exclusion to clients, the opposite of a public good (Callon 1994; Mirowski and Sent 2007). Mirowski and Sent (2007) have highlighted the problem of the commercialization and privatization of scientific knowledge. Furthermore, Mirowski (2009) avers that the commercialization of scientific knowledge is the apotheosis of a neoliberal program to promote and to construct free markets as the central condition of success of the neoliberal agenda. Mirowski argues that the problem of commercialization of scientific knowledge is that scientific knowledge is being transformed from a public good to a positional good. A further argument from Mirowski (2009b) is that science has been harmed by its commercialization.

Though Fuller's social epistemology is about the democratization of scientific knowledge, Fuller is also involved in the debate on whether scientific knowledge is a public good or a positional good. In addition, Fuller's social epistemology is about the governance of scientific knowledge through knowledge policy; it defends knowledge as a public good through his governance of science and the university as a site to defend against neoliberalism.

I thank Philip Mirowski and Jeroen Van Bouwel for comments on this paper. I also thank Steve Fuller and Philip Mirowski for their participation in a Vancouver 2006 *Social Sciences in Science* session on *Neoliberalism: The Hidden Politics of Science Studies?* A version of this paper was presented at the May 2006 *Canadian Society for the History and Philosophy of Science* conference and at *The Social Sciences and Democracy: A philosophy of science perspective* conference, 28–30 September 2006 in Ghent, Belgium.

This chapter examines Mirowski's and Fuller's views on the commercialization of scientific knowledge. I explicate Mirowski's views on science and society and Fuller's social epistemology. Then, I compare Mirowski's and Fuller's views.

Mirowski on science and society

Mirowski uses economic concepts to analyze how science is being commercialized. Some economic concepts used are the marketplace of ideas, rational choice, utility maximization and self-interest. Mirowski's major argument is that "the dominant structures of the funding and organization of scientific research in a particular culture *select* those doctrines which give appearance to rationalizing their practices" (Mirowski 2004, p. 322). Mirowski's view is that one starts with a description of research funding, organization and dissemination within a particular context in order to discuss how science does and does not 'work'; and that problems about conflict between science and society can be made more understandable from within this framework. In other words, Mirowski states that to understand the relationship between science and society, one should understand how science is funded and organized for this affects how science is produced. Mirowski extends his argument to Dewey, Reichenbach and Kitcher in that he avers that they rationalized the status quo in science. Mirowski notes that "it is argued that certain configurations of science organization, in conjunction with certain widely accepted images of society, have given rise to very specific orthodoxies in the philosophy of science: these represent the scientific dimensions of social knowledge" (Mirowski 2004, p. 289). Some of us may read that Mirowski is arguing for a causal relationship between funding and how science works, but he is not making that argument. Instead, he is arguing for functional interdependence of funding and scientific production. Functional interdependence is weaker than a linear causal relationship. It is similar to constant conjunction.[1]

The main theme of Mirowski's (2006; 2008) and Mirowski and Sent's (2007) papers is that scientific knowledge has been transformed from a public good and has been commercialized and globalized to a positional good. Mirowski argues that science is composed of a comprehensive view of the activities of science as part and parcel of a larger ensemble of actors. Mirowski focuses on the state, the corporate sector, the educational sector and 'civil society' agents such as foundations, NGOs, and citizen groups. They have been concerned with both the financial support and functional organization of scientific research in the past century. Mirowski's empirical thesis is that in the twentieth century, there have been three relatively distinct regimes of science funding and organization in the United States (the regimes rationalized by Dewey, Reichenbach and Kitcher, respectively), and each has involved a different alliance of state, corporation, university and foundation.

While commercialization is not new to science, distinct forms of commercialization have emerged in the contemporary era of globalization. Since 1980, a broad array of innovations initiated from within the corporate and state sectors have caused a profound reengineering of the university and of the structure of science. In the United States, several acts were passed that affected the organization of science. Mirowski and Sent focus on recent innovations in intellectual property, industrial R&D, multiple authorship in scientific journals and commercialization of the university. In 1980, the Bayh-Doyle Act was a major turning point in the United States, allowing "universities and small businesses to retain title to inventions made with federal R&D funding, and to negotiate exclusive licenses" (Mirowski and Sent 2007, p. 657).

According to Mirowski, neoliberals developed institutions, which included the Mont-Pelerin Society, Institute for Economic Analysis, the Hoover Institution and the American Enterprise Institute, in the social sciences especially in economics. Neoliberals were members of academic departments such as the Chicago economics department and law school. Inspired by Hayek, neoliberalism holds that the market is a better conveyor of information than the state and the "greatest information processor known to mankind" (Mirowski 2009). Following Hayek, Mirowski avers that neoliberalism advocates for the manufacture of ignorance. Man is ignorant and it is the market that has knowledge (Mirowski 2009c).

Fuller's social epistemology

Fuller on science and society

The purpose of Fuller's social epistemology is to use a social naturalistic theory of science to develop a rational knowledge policy for knowledge production. As a moral project of humanity, the social sciences connect with Fuller's social epistemology because its central issue is a normative project of how science should be organized, resulting in a knowledge policy consisting of democratically chosen goals (Fuller 2006, 2009a, 2009b; Remedios 2009).[2]

Fuller's political social epistemology holds that science and society interpenetrate each other. They are two separate forces that shape each other, but they are not separate entities. Fuller notes that:

> Science reproduces social tendencies in the sense that many features of science are simply features of the larger society. However, social forces can penetrate the processes of science more deeply, even transforming the content of knowledge. For example, as science is popularized, people attach their own meanings to scientific concepts and findings, which make it difficult for scientists to continue using the concepts and findings. The concept of 'gene' is a good example. (Fuller 1999, p. 278)

Fuller holds that science as an institution is itself a polity. This is not a new concept. Michael Polanyi (1962) articulated the notion of a republic of science and Thomas Kuhn and Karl Popper debated the issue of whether science represents a more 'closed' or 'open' society in 1965 at the London School of Economics (Lakatos and Musgrave 1970). Fuller agrees with Popper's Big Democracy notion of science, which claims that science is an institution that is ruled by scientists, who in turn form an elite. Fuller holds that science should be democratized to allow the public—who do not form an elite—to participate in making decisions about the institution of science. While Fuller does not provide justification for the superiority of democracy, he argues that science advertises itself as universally accessible and so to deny the democratic governance of science in practice is to admit that science has not fully realized itself (Fuller 2000b). On the normative level, Fuller stages an investigation regarding the nature of hypothetical norms. They include the regulation of the production and distribution of scientific texts (Schmaus 1991, p. 121).

Fuller on knowledge as materially embodied in society

Fuller's social epistemology does not directly address traditional problems of knowledge such as justified true belief because he is interested in how texts become certified as knowledge. He inquires into "the possible circulation patterns of these artifacts (especially how they are used to produce other such artifacts, as well as artifacts that have political and other cultural consequences); and the production of certain attitudes on the part of producers about the nature of the entire knowledge enterprise (such as the belief that it 'progresses')" (Fuller 1988, p. xii).

Fuller holds that knowledge is materially embodied within society. In keeping with naturalism, he argues that some outcomes of causal interactions between knowers and the world are embodied knowledge. Fuller's view is that embodied knowledge is a product, a commodity that costs time, effort and money to produce. Joseph Rouse notes that, for Fuller, knowledge is multiply embodied: "in specific texts, utterances, performances, and artifacts; in the cognitive capacities of persons; in their institutions and their norms, structures, and pathways of communication; in distributions of power, resources, and effective access to the settings in which knowledge is made and made authoritative; and in the apparatus and other materials comprising experimental or observational systems" (Rouse 1998, p. 98, cf. Fuller 1992; 1993, p. xv).

Fuller on scientific knowledge as a public good

As a commodity, knowledge is descriptively a positional good and normatively a public good for Fuller. "The quantity of the material is constant," Longino notes of Fuller (Longino 2002, p. 171). Fuller holds to the descriptive notion that knowledge is a positional good has value only in terms of its scarcity or restriction on its access, either through academic credentials or intellectual property rights (Fuller 2002, pp. xviii–xix; cf. Longino 2002, p. 171n45).

Fuller also adheres to the notion of the universality of scientific knowledge. Fuller defends the normative notion that scientific knowledge is a public good because scientific knowledge is a universal form of knowledge and hence access should be provided. If everyone has knowledge, then its value as a commodity is nil. Enlightenment thinkers consider universality to be a long-term consequence of extensive criticism of each other's attitudes, while Positivist thinkers hold that universality is located in the disciplined mental attitude of professional inquirers at the start of inquiry (Fuller 1997, p. 26). The notion that science is critical inquiry and that it should inform democracy has been held by democratic theorists from the American founding fathers through to John Stuart Mill, John Dewey, Karl Popper, and Jurgen Habermas (Fuller 1997, p. 4).

Fuller holds that the difference between Enlightenment and Positivist conceptions of science is that the former holds that contestation as a form of critical inquiry is important in order not to reproduce tradition mindlessly, while the latter holds that the disciplined mental attitude of professional inquirers is important in order not to have the chaos of disordered thought. The former holds that "science" as a mentality can potentially pervade all society and is not restricted to an elite group, while the latter holds that science is restricted to academic specialists, professional 'scientists', who administer to society as civil servants. Fuller notes that both movements are committed to universality as a feature of scientific knowledge, which makes knowledge, knowledge for everyone. Hence, both movements are committed to knowledge as a public good. Though Mirowski does not state that knowledge should be a public good, his view is that science has been harmed by the commercialization and globalization of knowledge.

Fuller's governance of science

With the advent of naturalistic epistemology and philosophy of science, there have been very few attempts to institute the governance of science or outline what governance might look like. Some naturalistic philosophers, such as Philip Kitcher (1993), Alvin Goldman (1999) and David Hull (1988), hold to an 'invisible hand' account of science, which is opposed to a governance account of science. To regulate science, Fuller has provided a governance account, which is a key notion in his social epistemology. Regulation of science amounts to governance of science instead of science being operated through the invisible hand of the market.

In suggesting governance, Fuller is recommending legislation in which a constitution is established to govern the community of scientists. Because science is a social institution, and science is also a systematic pursuit of knowledge that covers all disciplines, Fuller holds that science should be accountable to the democratic state. It is through a constitution that Fuller's hypothetical policy norms would have force on scientists. A constitution is a social contract meant to establish social order. Fuller (2000a) argues that only a republican political theory of science is compatible with an 'open society', or the sort of political society that Fuller (along with Popper) holds to be the normative vision of science. Fuller argues that because science is a universal form of knowledge that applies to everyone, all people deserve to play a role in directing it. A republican theory of science holds as its main principles a participatory form of politics and the right to be wrong. The principle of the right to be wrong is an articulation of Philip Pettit's theory of nondomination. Nondomination is the central concept of republicanism (Pettit 1997). In an open society that welcomes criticism, all citizens should be able to express their opinions freely without the threat of harmful consequences. According to Fuller, given the ways current scientific research and teaching are organized, individuals stand to lose their right to be wrong when the opportunities to advance bold conjectures are constrained by 'Big Science'. Big Science is a way to describe the material transformation of science into something that requires larger and larger investments—for example, the superconducting supercollider—leading to the inequality of power and money. The current realities of scientific investigation do not live up to the normative vision of the open society, and so, Fuller recommends changes in the way that science is governed.

Fuller provides recommendations for a redesigned institutional structure for the governance of science. The heart of Fuller's proposal is the secularization of science, in which basic research is institutionally separated from the state, and scientific knowledge does not have privileged authority within politics and the public sphere. This leads to a two-stage model for the reorganization of science that parallels Fuller's reinvention of the context of discovery and the context of justification. I suggest this is where Fuller addresses the neoliberal notion of science as the marketplace of ideas. The first stage of the model, which parallels the context of discovery, is that the state should stay out of basic research and discovery should be privately funded. This stage leads to a market system of corporations, interest groups and others that work to discover knowledge. This system would inevitably produce less new knowledge, and hence, would produce fewer unread publications. Hence, science is commercialized and globalized. In the second stage of the model, which parallels the context of justification, state-funded universities would test and distribute privately generated knowledge. This is how commercialized and globalized scientific knowledge is justified and its distribution managed. Scientific knowledge is transformed from a positional good, which is research done by private corporations, to a public good, which is distributed by universities.

There are three important steps in realizing Fuller's concrete policy measures, configured as elements of a constitution to democratize science. First, scientists should be required to defend their proposals to one another in an open forum to ensure that research is epistemically fungible, which is to say that no inquiry should have so much funding that its competitors are excluded. Second, there should be a cross-disciplinary assessment of research proposals. Third, the voice of the public should be strengthened with 'science elections'. Those elections may well result in the redistribution of research funds (Fuller 2000a, pp. 135–51).

Comparison of Fuller's and Mirowski's views

To compare Mirowski's and Fuller's positions, I use Fuller's framework of external and internal social epistemology (Fuller 2007) as a comparator. External social epistemology presumes a "common understanding of the current state of science but differs on the ultimate ends that science should serve, and hence how the present is projected into the future" (Fuller 2007, p. 146). For Fuller, external social epistemology is legislative knowledge policy. Fuller's example is the Mach/Planck debate on the ends of knowledge (Fuller 2007, pp. 146–9). I suggest that Fuller is both an external and an internal social epistemologist. As an external social epistemologist, Fuller discusses the Mach/Planck debate. An internal social epistemologist's concern is the degree to which scientific practices approximate the best regime (Fuller 2007, p. 146). Disagreements in internal social epistemology presume agreement on a set of ends but disagreement on the current empirical state of science. Fuller is concerned with the degree to which scientific practices approximate the best regime, which is a judicial knowledge policy that addresses epistemic justice issues such as research fraud (more discussion in the Conclusion).

In the previous section, 'Mirowski on science and society', I described how Mirowski treats science as a subsystem of a larger social system: Mirowski questions the role of philosophy of science as advocated by Kitcher (1993) and Longino (1990, 2002) in that they do not recognize how the organization of science and funding of science affect how science is produced. In this vein, Mirowski can be considered partially to be an external social epistemologist. As for disagreement over the ends of science, which Fuller considers to be a hallmark of external social epistemology, Mirowski does not provide an argument about what those ends should be. Mirowski's main argument is that science has been harmed by commercialization and neoliberalism advocates for the manufacture of ignorance, hence, neoliberalism should be revoked. In this vein, Mirowski cannot be considered as an external social epistemologist. As for internal social epistemology, Mirowski's difference from Fuller is that Mirowski does not regard that there is a best regime to which science approximates. Hence, Mirowski cannot be considered as an internal social epistemologist.

Though Mirowski does not have a theory on how science should be governed as Fuller does, how does Fuller's two-stage model measure against Mirowski's arguments against commercialization? I suggest that Mirowski may disagree with Fuller's first stage in which the state stays out of basic research and science is commercialized. Mirowski may also question Fuller's second stage in which scientific knowledge is distributed by universities, because, for Mirowski, the more universities are involved in market activities, the more state support may diminish so public universities may become privatized (Mirowski 2009c). Guston has raised a similar criticism that, during the Cold War, the U.S government funded 70 percent of scientific research and private industry funded 30 percent. Today, the funding has been reversed so the private industry funds 70 percent and the U.S. government funds 30 percent (Guston 2002, p. 199). Also, Mirowski may question how the gap is bridged in Fuller's two-stage model as it transforms commercialized science as a positional good into science as a public good distributed by universities.

Of the many other criticisms[3] of Fuller's notion of governance, my own main criticism is that if discovery were privately funded and the state did not regulate what is researched, then how would the state regulate against science that it finds morally reprehensible? An outstanding example is human cloning, a procedure that currently is banned in Canada. Moreover, if research were privately funded, then why would scientists agree to Fuller's three concrete policy measures (Remedios 2003, p. 81)? Fuller's answer to my criticism is that competitive space needs to be structured in terms of a constitution for science. Markets require background laws to be in place, if only to minimize transaction costs. A constitution for science provides the background conditions against which competition can occur. Research production is privatized in the public sphere in terms of science elections. On issues such as biotechnology, Fuller's "personal view is that research should never be stopped, but serious political thinking needs to be given to the distribution end, e.g., how it is taught, represented, sold, given away, etc. Research should only be stopped, if there is a better proposal to spend the money involved, not as a matter of principle" (Remedios 2003, p. 81).

Conclusion

Mirowski (2009a) notes that neoliberalism, as a constructivist doctrine, is achieved through intervention to create the good society rather than it come about naturally in the absence of political effort through laissez-faire capitalism. Though neoliberalism holds that the market may surpass the state in processing information, my response is that the market is not a better conveyor of knowledge because outsourced scientific organizations can provide inaccurate research, which makes the marketplace of ideas look like it is performing better than it is. The most recent example is the meltdown of the U.S. and global financial markets and the more than trillion-dollar bailout by the U.S. government and other governments. The lack of regulation by the United States and other governments of the financial markets allowed for conditions, which included fraud, that led to the collapse of big U.S. investment banks such as Bear Stearns and Lehman Brothers and banks in the United Kingdom and Iceland. Analogously, with the lack of regulation of science and the increase of globalization and commercialization, there is no guarantee that fraud can be detected adequately in science. Fuller's internal social epistemology is a judicial knowledge policy, which is about research ethics. Fuller asks if there is a need of national research ethics boards to address issues of research fraud. The two examples Fuller uses are the South Korean scientist Professor Wang, who misrepresented data of extracting stem cells from human embryos, and Bjorn Lomborg, the Danish social scientist who was investigated by the Danish government for his book *The Sceptical Environmentalist* (Fuller 2007, pp. 149–78). The answer to Fuller's question is the subject of another paper.

Notes

1. This is in contrast to the linear model discussed in the section titled 'The things that economists say about science' in Mirowski's contribution to this volume, 'Some Economists Rush to Rescue Science from Politics, Only to Discover in Their Haste, They Went to the Wrong Address.'
2. Cf. Fuller 1993; Remedios 2003.
3. Bowden (2001), Collier (2000, 2002), Guston (2002), Keith (2002), Lópéz Cerezo (2002), Radder (2000), Ravetz (2002) and Wray (2001).

References

Bowden, G. (2001) Review of *The Governance of Science* by Steve Fuller. *Canadian Journal of Sociology* 26(4), 675–8.

Callon, M. (1994) 'Is Science a Public Good?' *Science, Technology, & Human Values,* 19(4), 395–424.

Collier, James (2000) Review of *The Governance of Science* by Steve Fuller. *Minerva* 38(1), 109–20.

Collier, James (2002) 'Scripting the Radical Critique of Science: the Morrill Act and American Land-Grant University.' *Futures* 34, 182–91.

Fuller, S. (1988) *Social Epistemology.* Bloomington: Indiana University Press.

Fuller, S. (1992) 'Knowledge as Product and Property.' In: *Culture and Power of Knowledge*, ed. Nico Stehr and Richard Ericson. Berlin: Walter de Gruyter.

Fuller, S. (1993) *Philosophy of Science and Its Discontents*, 2nd ed. Boulder, CO: Westview Press.

Fuller, S. (1997) *Science.* Buckingham, UK: Open University Press.

Fuller, S. (1999) 'Response to Japanese Social Epistemologists: Some Ways Forward for the 21st Century.' *Social Epistemology* 13 (3/4), 273302.

Fuller, S. (2000a) *The Governance of Science: Ideology and the Future of the Open Society.* Buckingham, UK: Open University Press.

Fuller, S. (2000b) E-mail to the author, 22 February.

Fuller, S. (2002) *Social Epistemology*, 2nd ed. Bloomington: Indiana University Press.

Fuller, S. (2006) *The New Sociological Imagination*. London: Sage Publications.

Fuller, S. (2007) *New Frontiers in Science and Technology Studies*. Cambridge, UK: Polity Press.

Fuller, S. (2009a, April) 'In Search of the Sociological Foundations of Humanity.' *History of Human Sciences* 22(2), 138–45.

Fuller, S. (2009b) 'Humanity: The Always Already – Or Never To Be – Object of the Social Sciences.' This volume.

Goldman, A. (1999) *Knowledge in a Social World*. New York: Oxford University Press.

Greenberg, D. S. (2007) *Science for Sale*. Chicago: University of Chicago Press.

Guston, David H. (2002) 'Secularising Science? *Futures* 34, 197–99.

Hull, D. (1988) *Science as a Process*. Chicago: University of Chicago Press.

Keith, W. (2002) 'Good Questions in Search of Good Answers.' *Futures* 34, 178–81.

Kitcher, P. (1993) *The Advancement of Science: Science without Legend, Objectivity without Illusions*. Oxford: Oxford University Press.

Lakatos, I. and A. Musgrave, eds. (1970) *Criticism and the Growth of Knowledge*. Cambridge: Cambridge University Press.

Lópéz Cerezo, J. A. (2002) 'The Governance of Science from the Periphery, or a Look at Budapest from Santo Domingo.' *Futures* 34, 192–6.

Longino, H. E. (1990) *Science as Social Knowledge*. Princeton, NJ: Princeton University Press.

Longino, H. E. (2002) *The Fate of Knowledge*. Princeton, NJ: Princeton University Press.

Mirowski, P. (2004) 'The Scientific Dimensions of Social Knowledge and Their Distant Echoes in 20th-Century American Philosophy of Science.' *Studies in History and Philosophy of Science* 35(2), 283–326.

Mirowski, P. (2009) 'Viridiana Jones and the Temple of Mammon; Or, Adventures in Neoliberal Studies.' In: forthcoming book *ScienceMart.*

Mirowski, P. (2009a) 'Defining Neoliberalism.' In: *The Road to Mont Pelerin*, ed. Philip Mirowski and Dieter Plewhe. Cambridge, MA: Harvard University Press, pp. 417–455.

Mirowski, P. (2009b) 'Has Science Been "Harmed" by the Modern Commercial Regime?" In: forthcoming book *ScienceMart.*

Mirowski, P. (2009c) 'The New Production of Ignorance: The Dirty Little Secret of the New Knowledge Economy.' In: forthcoming book *ScienceMart.*

Mirowski, P. and Sent, E.-M. (2007) 'The Commercialization of Science, and the Response of STS.' In: *The Handbook of Science & Technology Studies*, 3rd ed., ed. Edward J. Hackett, Olga Amsterdamska, Michael Lynch and Judy Wajcman. Cambridge, MA: MIT Press, pp. 635–90.

Pettit, P. (1997) *Republicanism*. Oxford: Oxford University Press.

Polanyi, M. (1962) 'The Republic of Science: Its Political and Economic Theory.' *Minerva* 1, 54–73.

Radder, H. (2000) Review of *The Governance of Science* by Steve Fuller. *Science, Technology & Human Values* 25(4), 520–7.

Ravetz, Jerry. (2002) 'The Challenge Beyond Orthodox Science.' *Futures* 34, 200–3.

Remedios, F. (2003) *Legitimizing Scientific Knowledge.* Lanham, MD. Lexington Books.

Remedios, F. (2009, April) 'Fuller's Project of Humanity: Social Sciences or Sociobiology?' *History of Human Sciences* 22(2), 115–20.

Rouse, J. (1998) 'New Philosophies of Science in North America—Twenty Years Later.' *Journal for General Philosophy of Science* 29, 71–122.

Schmaus, W. (1991, March) 'Review of Social Epistemology by Steve Fuller.' *Philosophy of Social Sciences* 21(1), 121–5.

Wray, K. Brad. (2001) Review of The Governance of Science by Steve Fuller. International Studies in Philosophy of Science 15(1), 110–12.

12
Humanity: The Always Already – Or Never to Be – Object of the Social Sciences?

Steve Fuller

Is the success of the social sciences predicated on our humanity?

From one viewpoint, the social sciences have never been more successful, especially in terms of available research funding and student course demand. Moreover, certain social science methodologies, notably those related to game theory, rational choice theory, and actor-network theory have been used to model phenomena in the life sciences. This would suggest that the social sciences are extending their influence across disciplinary boundaries. However, at the same time, 'social science' is losing its salience as a brand name or market attractor. In more academic terms, the social sciences are losing their distinctiveness as a body of knowledge distinguishable from, on the one hand, the humanities and, on the other, the natural sciences. That distinction was epitomised in the idea of a 'universal humanity' as both a scientific object and a political project that was explicitly developed by Christianity's most faithful secular offspring – German Idealism, French Positivism, and the Socialist movements of the nineteenth and twentieth centuries. Each challenged, on the one hand, the humanities by declaring equal interest in all of humanity (not only the elite contributors to the 'classics') and, on the other, the natural sciences by declaring a specific interest in humans (in terms of whom other beings are treated as a secondary consideration, if not outright means to human ends). This chapter should be understood as an extension of Fuller (2006), a call to revive this robust sense of social science under the rubric of a 'new sociological imagination'.

However, the call faces an uphill struggle, as was amply documented in Baber (2009), a symposium devoted to Fuller (2006). As against my 'all and only humans' approach of the social sciences, the humanities and natural sciences are rediscovering their common historic interest in *human nature*, with stress now unequivocally placed on the 'nature' rather than the 'human'. In the face of the social sciences' tendency to attenuate if not outright reject human nature over the past quarter-millennium, the past 30 years have witnessed a steady stream of works purporting to 'unify knowledge', most explicitly Wilson (1998), which in practice would make direct links between the classical humanities and the modern natural sciences by circumventing the social sciences altogether in the name of 'human nature'. Here Darwin replaces Aristotle as the grand unifier. In this context, the concept of universal humanity and most social science *theories* (though, as I have already suggested, not social science methods or findings) appear as vestiges of a monotheistic worldview that would elevate the human condition above the rest of nature.

Notwithstanding the radically different biologies that underwrote their conceptions of human nature, Aristotle and Darwin both doubted that the traits most closely associated with normative conceptions of 'humanity' were equally distributed across all members of *Homo sapiens*. Whereas Aristotle and his contemporaries argued about the limits of pedagogy in converting the upright ape into a political animal, Darwin and his successors have suspected that the upright ape's various attempts to transcend its biological condition – be it via Christianity or Socialism – simply reflects a pathology in an overdeveloped cerebral cortex.

Moreover, the general prognosis of the re-absorption, if not outright 'withering away', of the social sciences into a broader conception of nature has been also advanced by a consensus of postmodern social theorists who have queried the ontological significance of the human/nonhuman distinction and the need for disciplinary boundaries altogether (Latour 1993; Wallerstein 1996). However, their antidualism is informed less by a desire to reduce the mental to the physical than to reveal the interpenetration of the two categories, such that spirituality or consciousness is no longer seen as unique to humans but common to even the simplest forms of matter. This is not behaviourism or even materialism, at least as conventionally understood by physics-minded philosophers, but something closer to hylozoism and even panpsychism (e.g., Deleuze and Guattari 1987). Such convergence between naturalists and postmodernists should be unsurprising, given their common basis in Darwin's explicitly nonteleological version of evolutionary theory. In the postmodern case, it is filtered through Nietzsche's 'genealogical method'. The benchmark text here is Michel Foucault's *The Order of Things* (1970), his most sustained 'archaeology of knowledge', which focused on the sudden emergence and gradual disappearance of the object 'man', that 'empirical-transcendental doublet', the Kantian phrase that Foucault used to characterise the distinctive nature of our being. In the Foucaultian gaze, we are exotic apes suffering from what Richard Dawkins (2006) calls the 'God delusion'.

Foucault notoriously regarded humanity as a historically bounded object that really only came into existence with Kant's coinage of 'anthropology' in 1798 when he was addressing how beings of such diverse racial-cultural histories as those Linnaeus had canonised less than 50 years earlier as

'*Homo sapiens*' could ever deliver on the Enlightenment promise of 'world-citizens' – or 'cosmopolitans', to recall the original Greek (cf. Toulmin 1990). Kant's cosmopolitan conundrum slowly began to lose its salience a hundred years later, as Marx, Nietzsche and Freud, each in his own way, portrayed the human as an unstable compound, a 'house divided against itself' subject to false consciousness, self-deception, and/or repression. For them 'humanity' in this grandiose sense merely encouraged people to live in an unrealisable future that diverted them from the intractable problems they currently faced.

Contrary to most of Foucault's critics, I accept the *prima facie* plausibility of his radically demystified account of the concept of humanity, which in turn demands a systematic response, one begun in Fuller (2006). Foucault is certainly correct that a distinct body of knowledge called the 'human sciences' or 'moral sciences' or 'social sciences' that takes all human beings to be of equal epistemic interest and moral concern has been most compelling from the late eighteenth to the late twentieth century. For Foucault himself, this was a blip on the radar of Western intellectual history, on either side of which he espied (before) an enchanted and (after) a disenchanted naturalism: in short, Aristotle and Darwin. Even those operating within a more conventional view of intellectual history can recognize Foucault's 'Age of Man' as signifying the shift from a broadly supernaturalist to a broadly naturalist worldview: For example, where once wars were fought about the right approach to God (theology), wars in the future are likely to be fought on the right approach to nature (ecology), with the familiar modern inter-state conflicts licensed by the Peace of Westphalia of 1648 functioning as an extended transitional phase between the two pure forms (sociology).

The precariousness of the human: Why Foucault is (unfortunately) correct

As propaedeutic to my own response, we need to get the full sense of humanity's ontological precariousness to appreciate the depth of Foucault's challenge. It is epitomized in the following question: *Have we always, sometimes or never been human?* The more one understands the history of the concept of humanity, the less the question appears frivolous. To take the question seriously, one should take into account the following four considerations:

1. There has always been ambiguity about where to draw the line between humans and nonhumans (Corbey 2005). It can be found even in Linnaeus's coinage of our species name *Homo sapiens* in the mid-eighteenth century. At the level of morphology, Linnaeus did not see a sharp difference between the higher order apes and the various human races. However, as a special creationist of the Lutheran persuasion, Linnaeus believed the biblical claim that all humans were endowed with souls that gave them the potential to hear God's call – even very fallen humans, such as the sons of Ham cursed by Noah, from whom Africans were thought to descend. In this respect, the Bible made up for the shortcomings of empirical observation in providing a clear definition of the human (Koerner 1999). It is only a short step from this line of reasoning to the 'standard of civilization' long enshrined in international law – that a people are properly 'human' only if they heed the call of God or at least, in more secular terms, tolerate the commerce of those who do (Fuller 1997, chap. 7). The behaviourist orientation of this approach is striking. The distinctive spirituality of the human is marked by one's responsiveness to a sacred book in which the distinction is itself inscribed. Thus, one reason why so many more American Indians than Black slaves were slaughtered in the United States was that the former refused to adjust their mode of being in response to the divine call.

2. *There has always been recognition of the diversity of physical and mental qualities of beings that might qualify as humans.* Sometimes the originality of this observation is credited to Darwin but only because folk notions of species tend to presume a crude understanding of essentialism. Even Aristotle knew that a species contains differentia: The same thing may exist in many different ways, amongst which exists what Wittgenstein called a 'family resemblance' that, in turn, points to a common ancestry. Where Aristotle and Darwin disagreed was that the former thought of this variation as resulting from a mixture of elements provided by the particular parents whereas the latter saw it as endemic to the general process of reproducing the species. Nevertheless, followers of both Aristotle and Darwin have had their doubts about the capacity of all members of *Homo sapiens* to achieve the same levels of humanity. To be sure, they believed that all members of the species possess a sufficiently joined up nervous system to merit the minimal infliction of gratuitous pain. But otherwise, people are inherently different, which means that the just society is organized by enabling each person to flourish in the sort of life that he or she has been designed to lead. In this respect, the division of labour is simply the outward sociological expression of a natural biological tendency, which (so at least Plato believed) philosophy could rationalize.

3. *There has always been an understanding that not everything about humans makes one human and that one's humanity might be improved by increasing or decreasing some of its natural properties.* In other words, to be human is to engage in activities whose purpose goes beyond the simple promotion and maintenance of the animal natures of those qualified to be human. In Western philosophy and theology, one normally characterizes such matters as involving a 'spiritual' or 'intellectual' quest, but it is perhaps less misleadingly cast as a call to artifice. Here we might identify three 'Ages of Artifice'. (1) *The Ancient Artifice*, epitomized by the Greek ideal of *paideia*, instilled humanity through instruction on how to orient one's mind and comport one's body to justify one's existence to others as worthy of recognition and

respect. In practice, this meant speaking and observing well – the source of the liberal arts disciplines. (2) *The Medieval Artifice*, epitomized by the introduction of *universitas* into Roman law, promoted humanity by defining collective projects into whose interests individuals are 'incorporated', say, by joining a city, guild, church, monastic order or university. Here one exchanges a family-based identity for an identity whose significance transcends not only one's biological heritage but also one's own life. (3) *The Modern Artifice*, epitomized by the rise of engineering as a distinct profession, advances the human condition by redesigning the natural world – including our natural bodies – to enable the efficient expression of what we most value in ourselves.

4. *There has always been recognition that genuine humanity is precious and elusive, and hence 'projects' and 'disciplines' for its promotion and maintenance have been necessary.* In a sense, this is the negative side of humanity's inherent artificiality, noted above. It implies that the pursuit of humanity may not necessarily serve the interests of all flesh-and-blood humans. At the very least, not all humans may benefit to the same extent and in the same way from the process of 'humanization'. The easiest way to appreciate this point is to consider what it would take to realize any of the historically proposed schemes that would establish 'equality' among all humans. Some individual humans would be raised and others diminished in the process, the balance between which would always need to be monitored. Christianity is largely responsible for inducing widespread cultural guilt about the failure of all members of *Homo sapiens* to be treated as humans, which in turn opened a long and ongoing discussion about how 'human potential ' (aka soul) might best be realized. However, it is only in the late eighteenth century that the first systematic efforts to raise 'the overall level of humanity' by the redistribution of wealth and sentiment are instituted, this time in the name of 'Enlightenment' (Fleischacker 2004). In the nineteenth and twentieth centuries, these efforts came to be routinized as a set of political expectations concerning mass education, health care and welfare provision more generally.

The perceived failure, or at least underachievement, in securing the fourth sense of humanity's ontological precariousness has led even self-avowed members of the political left to judge 'humanity' a fantasy whose inherent risks are outweighed only by its manifest hypocrisy. Often Foucault's 'death of man' thesis underwrites this conclusion, but as suggested by the five critiques listed below, this antihumanism can be found in a variety of contemporary trends, some of which stray far beyond Foucault's original concerns but all of which are well represented in John Gray's (2002) *Straw Dogs*, the most provocative British book of political theory in recent times:

1. *The Postmodern Academic critique*: 'Humanity' is a mask that hegemonic male elites don to exert power over everyone – and everything – else.

2. *The Neo-liberal (and Neo-conservative) critique*: Humanity as a political project costs too much and delivers too little (aka race and gender are 'really real').
3. *The Ecological critique*: The projects associated with humanity are depleting natural resources, if not endangering the entire biosphere.
4. *The Animal Rights critique*: Humanity's self-privileging is based on pre-Darwinian theological ideas that cause other creatures needless suffering.
5. *The Posthumanist critique*: Not even humans want to associate with other humans any more – they prefer other animals and the 'second selves', or avatars, they can create on their computers.

Humanity's perennial precariousness may be appreciated upon considering that prior to the Stoics the classical philosophers probably did not count all members of the species *Homo sapiens* as 'human' in a normatively robust sense, whereby *'Homo sapiens'* names only humanity's contingent biological starting point but not its ultimate realization. On the one hand, when Aristotle defined the human as *zoon politikon* ('political animal'), he seemed to be referring only to those with the capacity to participate in public life, that is, male landholders in good social standing. On the other, a quarter-millennium later and under the rubric of *humanitas*, the Stoic Cicero commended a variety of orientations to the world that transcend ordinary brute survival. They included the cultivation of leisure as an end in itself (and not simply a respite between periods of work) and the recognition of both what others accomplish on their own and one's own dependency on others (thereby evening out the natural tendency towards pure self-interest). For Cicero, himself a semi-invalid provincial who eventually achieved greatness in the Roman Senate and as a writer of Latin prose, *humanitas* helped to explain his own success. Indeed, Cicero's contemporaries deemed him a *novus homo*, a 'self-made man', the ultimate compliment that could be paid to a being so marked by artifice. In this respect, *pedagogy*, the ancient discipline that grew out of rhetoric and preceded government as the means for radically and systematically amending the upright ape's default tendencies, should be seen as the low-tech precursor of the various treatments increasingly available today for 'cognitive enhancement' (cf. Ingold 1994; Harris 2007).

Here it is important to recall that in the classical world, the default social unit was the family estate, or household (*oikos*), and *not* the city-state (*polis*). The latter only came into its own under extraordinary conditions, either in times that require mutual aid (i.e., in war or a famine) or when the material bases of life have been already served (i.e., in leisure). In contrast, the household was the natural habitat of several families of human and non-human species (i.e., farm animals) that have long coexisted symbiotically. It involved a functional differentiation of resource production and management based on the workings of 'natural justice', that is, the spontaneous variation in individual talents within a species. Just as one would not expect everyone to be equally capable of hard physical labour, the same would be true of the mental discipline necessary for becoming 'human' in the normatively robust sense indicated above. In this sense, a 'just society' removes any artifice that might prevent heredity from operating as an efficient sorting mechanism for assigning individuals to their appropriate societal functions. Thus, the less articulate would not feel the burden of having to speak 'rationally' because they would be valued for their other natural capacities. In short, the 'just society' did not denote a vehicle for collective self-improvement – as that phrase would come to mean in the modern era – but a sustainable ecology of mutually complementary individuals.

In this context, Aristotle was more trusting than Plato of natural justice, evidence for which can be found in Aristotle's rather charitable view of Athenian drama, which tended to feature plots in which one or more characters tries to act contrary to nature, only to fail in some comic or tragic way. A good contemporary exemplar of the Aristotelian attitude is the controversial U.S. political scientist Charles Murray (2003), who has never ceased to find the comic and tragic elements in state-based welfare schemes. As for Plato, while he appreciated the persuasive force of hereditary appeals, especially as a socially stabilising ideology, in the end his scheme to recruit and train philosopher-kings was about finding the best individual for the job based on rational criteria, which justified a policy of artificial selection. From his standpoint, the Athenian dramatists ran interference on humanity's capacity to realize nature's ends more fully. However, when it came to the city-state as a society of self-legislating equals, both Plato and Aristotle found it an alluring but potentially self-destructive chimera.

What Plato and Aristotle lacked was a criterion of humanity that overcame the obvious morphological and behavioural differences among members of *Homo sapiens*. A century later, to the rescue came the rather complex Stoic idea of the soul, which encompassed words like *pneuma*, *psyche*, *conatus* and, most notably, *logos*. These words captured the source, the expression and the perpetuation of life. By today's standards the nature of these entities blurred distinctions between the psychological, biological and physical. Each was located somewhere between a meme and a gene, the sort of thing that only a Lamarckian could truly love. But together they served to shift the burden of proof to a recognisably universalistic notion of humanity, whereby instead of marvelling how well certain people can speak in public (the canonical expression of *logos*), one wondered why everyone else *cannot* do so as well – given that God had also endowed them with a soul.

Perhaps Stoicism's most enduring legacy to the concept of humanity has been the Christian gloss on *logos* as divine agency in the Gospel of John. However, John radically shifted the metaphysical horizon of *logos*. The Stoics regarded humans as embedded in a pantheistic universe: Our capacity to resist the animal passions and to reason beyond our immediate needs reflected our unique status as a microcosm of the universe – but nothing more. In other words, as Spinoza continued to believe, humanity was simply the locus of God's self-understanding, where 'God' is simply a pious name for nature in its entirety. In contrast, the Christian God unequivocally transcends the world of his creation, and humans are defined as those created 'in his image and likeness', what after St Augustine has come to be known as the *imago dei* doctrine. The difference between Catholic and Protestant sensibilities turns on what one takes to be the main feature of the *imago dei* doctrine: *our subordination* or *our likeness* to God – what in the next section I characterise as, respectively, the 'Paris' and 'Oxford' spin on the doctrine.

Humanity as a bipolar disorder and the legacy of John Duns Scotus

Foucault understood well the latent source of antihumanism in the Western tradition. Western theology poses the question of humanity in terms of whether we are more like gods or apes. From a sociological standpoint, either answer ends up devaluing what most normal human beings do, or at least what they believe about what they do. Those who would urge humanity's apotheosis are eager to discipline, replace, if not outright eliminate our animal natures to release a frictionless medium, typically of thought, that enables us to merge with God. Their sense of science is ascetic, such that experiments function as trials of the soul, where the inquirer is pressed into extreme situations that challenge our physical senses to elicit a significant response (Noble 1997). The flagship discipline of apotheosis is *optics*, whose imprint is still felt in the hype surrounding superconductivity research and the eagerness with which people embrace avatars in cyberspace and speculative attempts to download consciousness into silicon chips. In contrast, those stressing our ontological proximity to the apes have tried to show the continuities in our natural modes of being with those of the rest of the animal kingdom, typically to parlay a respect for nature into a sense of humility, if not submission. Their sense of science involves full sensory immersion in a habitat, the flagship discipline of which is *natural history*. From this standpoint, Darwinists overstep the line of theological respectability *only* when, as in the case of the 'new atheist' followers of Dawkins (2006), they infer the nonexistence of a supernatural realm simply because it cannot be accessed through science: Rather, as the Roman Catholic Church has stressed since Thomas Aquinas, natural scientists should understand that other modes of being require other modes of knowing.

For today's version of the 'gods' versus 'apes' poles that pulls apart the integrity of humanity, consider Ray Kurzweil's (1999) 'spiritual machines' and Peter Singer's (1999) 'animal liberation' as radically alternative 'posthuman' ends – that is, what humans should be about, understood in terms of the larger reference group with which we wish to identify in the future. Speaking of 'human' as what Goodman (1954) would call a 'projectible predicate', we see here what I have called the great *carbon-silicon divide* in human being (Fuller 2007b: chap. 2). It is epitomized in the following question, which brings out the alternative modalities at stake: *Are we by nature intellects that happen for now to possess animal bodies (Kurzweil), or animals that happen for now to possess distinctive minds (Singer)?* The former option suggests a purposeful intelligence who explores different media for optimal self-expression, each disposable if proved unfit for purpose. The latter option implies a completely contingent process that reduces any noteworthy effects to emergent properties of particular combinations of elements, the valorisation of which is superstitious. Theologically speaking, Kurzweil and Singer are guilty of complementary excesses that recall the Gnostics and the pagans, respectively, as boundary challengers to Christianity from its earliest days. Kurzweil the Gnostic would 'sacrifice' his own body – and perhaps that of others – in service of immortal life, while Singer the pagan would 'sacrifice' his higher mental functions in the sense of removing the inhibitions they normally pose to a full return to our sensuous, mortal roots.

Notwithstanding the contemporary focus of their interests, Singer and Kurzweil are reproducing signature attitudes towards the ends of knowledge that are traceable to the university foundations of Paris and Oxford, respectively, in the mid-thirteenth century, and their contrasting views of the ends of humanity. This 'bipolar disorder' is captured in Table 12.1, which is informed by the excellent treatment of John Duns Scotus presented in Williams (2007). (See also Fuller 2008: Chapter 2.)

John Duns Scotus (1266–1308) is central to this discussion because *Homo sapiens* embarked on humanity as a collective project of indefinite duration only once an appropriate metaphysical framework was in place to make good on the biblical idea that humans have been created 'in the image and likeness of God'. Although St Augustine first crystallised *imago dei* as a theological doctrine, it really only comes into its own a millennium later, with Duns Scotus, largely in response to the Thomistic tendency to multiply realms of knowing and being, seemingly to check the ambitions of the will as the locus of humanity's God-given creativity. But it took another quarter-millennium before the nonconformist Christians active in Europe's Scientific Revolution interpreted the Scotist doctrine of the 'univocity of being' – that divine qualities exceed human ones only in degree but not kind – to imply that we might not only understand and improve our animal existence but that we might even chart a path to divinity (Funkenstein 1986, Harrison 1998). 'Humanity' thus became the name of the project by which *Homo sapiens* as an entire species engaged in this ontological self-transformation. The idea implied here, *achievable perfection*, is the source of modern notions of progress, ranging from scientific realist attitudes that 'the true' is something on which all sincere inquirers ultimately converge (regardless of theoretical

Table 12.1 Humanity as a bipolar disorder: Paris vs. Oxford

University	Paris	Oxford
Academic exemplars	Albertus Magnus, Thomas Aquinas	Roger Bacon, John Duns Scotus
Religious order	Dominican	Franciscan
Philosophical anchor	Aristotle	Plato
God is...	Simple	Infinite
'Being' is...	Equivocal	Univocal
Key ontological divide is...	Between God and creatures	Between humans and animals
Humans are...	High-grade creatures	Low-grade creators
God-human relationship	God and we are 'simple' in different senses, aka logical vs. psychological simplicity	We are 'finite' and God is 'infinite' in the same sense, i.e., as indefinite extension
Divine attributes....	Are endlessly additive but no convergence because our language cannot refer to God	Converge at the limit because our language (as *logos*) ultimately refers to God
A priori knowledge?	Only to the extent that God grants it	It is how we always participate in divine mind
Epistemic certainty	We can achieve it only in the natural world	We can achieve it on our own without God's help
Will is...	Intellectual appetite	Creative source of being
Will is based on ...	*Affectio commodi* ('sense of advantage' – what best suits our interests)	*Affectio iustitiae* ('sense of justice' – what the part looks like from an imagined divine whole
Systems perspective	Local adaptation	Global optimisation
We are free to decide...	Whether to do what is (already) right or wrong	Which path we take, which may have good or bad consequences
Basis for ethics	To live a flourishing life	To explore human potential to the fullest
What is possible?	Whatever is probable (i.e., an empirical notion)	Whatever is conceivable (i.e., a semantic notion)
What is progress?	Increasing differentiation, complexification	Increasing purification, demystification

starting point) to more general notions of progress, whereby 'the true' itself is held to converge ultimately with 'the good', 'the just', 'the beautiful', and so on in some utopian social order (Passmore 1970).

To appreciate how the Scotist shift of humanity's centre of gravity from animal to deity triggers a modernist mode of humanity, consider a standard way of thinking about God as existing outside of space and time – that is, *sub specie aeternitatis*. Thus, Newton explicated divine omniscience as a function of God being equidistant from all times and places. What appears to change within space and time appears from the 'view from nowhere' as always equally present (Nagel 1986). A measure of just how far short humans normally fall of this standard of divine knowing is our default asymmetrical attitudes towards the past and the future: We ordinarily believe that the past, even if not knowable in fact, is at least fixed in principle, while the future remains unknowable in principle because it has yet to be fixed in fact. What modernists call 'traditional', 'conservative' or 'pre-modern' societies, experience this asymmetry as especially significant: Traditionalists invoke the past to enforce the legitimacy of the present, while they portray the future as hazardous if it breaks definitively with the past. The modernist imperative, in contrast, has been to redress such asymmetrical attitudes towards past and future, so as to bring it closer to God's point of view, whereby what we call 'past' and 'future' are equally knowable – and (to God) known. In effect, modernity has tried to simulate the divine standpoint from a human position by arguing that our knowledge of the past is not so secure and our knowledge of the future not so insecure, which in its own fallibilistic way approximates Newton's divine ideal of being epistemically equidistant from all moments in space-time.

Perhaps the easiest way to see this strategy at work is to look at the dual movement of the modern scientific attitude towards temporal affairs. On the one hand, modern science is sceptical towards received views of the past, not least because of science's extended sense of time's backward reach, which serves to cast doubt on both the constancy and the reliability of the information that can be gathered about alleged events now thought to be so far from the present. This is the basis of the critical-historical method that was the hallmark of the 'humanities' once it started to name a set of academic disciplines in the early nineteenth century. It also undercut the basis for traditionalism, which was increasingly seen as a mythical construct. On the other hand, modern science takes a more positive view towards our capacity to know and even control the future, especially through predictive experiments. Here the idea that the 'end of time' is not near but in the indefinite future allows the prospect of collective learning through a controlled process of trial and error, what Karl Popper (1957) famously called 'piecemeal social engineering'. Thus, the future need not be faced with foreboding and fatalism but with openness to change, since there is time to make and correct mistakes, and indeed to refine our sense of long-term planning.

This is what is typically meant by positive references to a 'natural science' attitude to human affairs that is already present in the works of 'utopian socialism'.

At a metaphysical level, the strategy to simulate a divine symmetry in humanity's attitudes to the past and future reflects the Scotist shift in the default sense of what is 'possible' from *the probable* to *the conceivable* (Fuller 2002). This, in turn, had a knock-on effect on the understanding of human agency. On the one hand, because the past was no longer treated as such a secure guide to the future, the current generation of humans had to take personal responsibility for what to do – they could not simply defer uncritically to dead ancestors. On the other, because the future was no longer seen as so threatening, the current generation could play with alternative courses of action, perhaps comforted by the thought that future generations, with the benefit of both hindsight and insight (if not foresight), might be able to cope with the consequences better than those whose actions generated them. An important transitional figure was the British Unitarian preacher, experimental chemist and confidant of the U.S. founding fathers, Joseph Priestley (1733–1804), who explicitly regarded scientific progress as the expression of divine providence (Passmore 1970: chap. 10). The totality of Priestley's corpus wedded a deep but deconstructive approach to the history of Christianity to an equally deep and productive commitment to experimental science and utopian politics (Johnson 2008). Arguably he was the first fully embodied 'modern man'.

Of course, it would be a mistake to see the project of humanity as the straightforward outcome of Duns Scotus's subtle logical analysis of being. Nevertheless, Scotus sowed its seeds when he introduced the modern abstract conception of law as something universally binding and equally accessible. He did this by a linguistic innovation that effectively divested the force of law of its divine origins. As Brague (2007) observes, in Latin the shift was signified by the replacement of the proper name *pater omnipotens* ('almighty father') with the generic attribute *omnipotentia* ('omnipotence'). Social life had to be reconceptualised and reorganized in specific ways to reflect this change, the results of which we continue to take for granted. I earlier referred to the 'Medieval Artifice' known as the *universitas*, whereby individuals would be legally incorporated into larger social wholes, like cities and universities, on a nonbiological basis. Scotus's linguistic innovation was accompanied by a sensibility that what previously had been concentrated in God for eternity, and perhaps harnessed by his papal or royal representatives on Earth for a limited time, could be instead distributed over an indefinite period as an increasingly secular project in which many people might participate. Democracy as a universal ideal was thus born of a devolution of the 'divine corporation' to many self-sustaining *universitates* (Schneewind 1984).

Indeed, rather than reflecting any divine deficiency, this process of ontological devolution reflected an 'optimisation' strategy, whereby God's will is realized by multiple means, the most efficient of which may be indirect, say, as executed by humans acting out of their own accord, which would allow us to be divine creatures and autonomous agents at once. 'Principal-agent', an expression that political scientists and economists nowadays use for the relationship between, say, the people and their elected representatives or a firm's shareholders and its board of directors, captures this newfound sense of God's reliance on humans as his 'agents', the paradigm case of which may be Jesus. This entire way of thinking about divine causality in nature was indicative of the controversial branch of early modern theology, *theodicy*, which tried to infer the principles of divine justice from a world that is supposedly 'the best possible' yet admittedly full of imperfection and even room for improvement. Theodicy attempted to exploit the obverse of the Scotist thesis that human virtues are diminished versions of divine ones: namely, that God can be understood as a rational calculator with superhuman powers. In that case, crucial to the collective instruction of humanity is the reverse-engineering of nature to learn of its divinely inspired 'functions', so that we are in a position to devise means to provide for nature's ends more efficiently. This mentality, which informed Linnaeus's original classification of species, nowadays goes under the secular name of 'systems-theoretic', where it has remained a powerful heuristic not only in biology but also across the sociotechnical sciences, including engineering, cybernetics, economics and of course structural-functionalist sociology (Heims 1991).

In this context, the criteria used for incorporation into such a 'system' – say, via the passing of an examination, the pledging of an oath, the payment of a fee – served as a constant reminder of the sense in which the social order is an open projection: A common past can be extended into a variety of futures, depending on exactly who is selected to reproduce the *universitas*. Unlike a royal dynasty or a caste system, conjugal relations need not provide the default mode of succession: Identity is literally always under construction and potentially available to anyone who can pass the membership criteria, regardless of origins. This idea acquired great force and generality in the modern era through *social contract theory*, an attempt to bridge sacred and secular history in a philosophical discourse inspired by the various mutually binding agreements, or 'covenants', through which God ensured the future of the Chosen People in the Old Testament. In the first instance, it implied a substantial overlap in the divine and the human intellects – certainly enough to allow for the sophisticated transactions described in the Bible that still intrigue game theorists (Brams 2002). But equally, it suggested that while any human project may be of indefinite duration, its continuation requires periodic rededication, as exemplified by the dialectic of elections and revolutions that defines the history of modern politics. That dialectic speaks to the centrality of what Max Weber would have recognised as *decisionism*, a modern remnant of the Scotist idea that, in matters of ethics and politics, humans partake of divine creativity by the sheer capacity to discount the past and declare one's decision to be the source of all that follows. (In philosophy of science, this view – of similar vintage – is called *conventionalism*.)

Renaturalisation as dehumanisation: The long march back to Scotland

Karl Marx exemplified the bipolar disorder that I have identified with the concept of humanity. On the one hand, his projection of a planned economy in the postcapitalist world would seem to count him as an extreme decisionist. On the other hand, the heightened state of rational will in this Communist utopia marks a radical break with the normal run of human affairs, which he famously characterised as a matter of people making their own history but not as they choose. By this Marx did not mean to suggest that someone (or something) else made it for them, though that might have been a reasonable conclusion to draw from the philosophical historiography of Hegel or any of his precursors in the Enlightenment or those who professed theodicy. Rather, Marx was pointing to a subtle distinction that remains inscribed in the asymmetrical attitudes towards decisions in ethics and epistemology, at least in analytic philosophy: *We decide what to do but not what to believe.* The overwhelming significance of unintended consequences in history – that the world does not turn out as we plan – underwrites this asymmetry. However, the asymmetry is by no means absolute, and in fact is really only obvious when the relevant beliefs are clearly evidence-led (Fuller 2003, chap. 11). In contrast, a strong sense of epistemic decision-making is integral to both Protestant theology and positivist accounts of scientific theory choice. Indeed, a broad church of philosophers over the past hundred years, from Nietzsche and James to Carnap and Popper, have asserted the efficacy of decisions by stressing our world-making capacities, which amount to our willingness to follow through on the basis of a set of self-legislated principles, even in the face of some, if not all, empirical obstacles.

Decisionism in this broad sense keeps with the spirit of the two founding figures of modern ethics, Kant and Bentham, who cast their own theories as being about *legislation* – in the former case based on the reasonableness of allowing everyone to take the same decision, in the latter case based on overall consequences for those affected by one's decision. After Schneewind (1997), it is now accepted that this signature feature of modern ethics – whereby one adopts the standpoint of the rule-maker, as opposed to the rule-follower or, still more passively, a being to which rules are applied – is an outcome of the secularisation of theodicy, whereby the sense of cosmic justice pursued by the deity in Malebranche and Leibniz was downsized during the Enlightenment into, respectively, deontological and consequentialist normative ideals of human action, again with Priestley playing a pivotal role in the latter case (Nadler 2008). What then began to slip away, already in the mid-nineteenth century, was the autonomy of the human decision-maker, a point to which the Foucaultian historiography of the social sciences has been especially sensitive.

On the one hand, the social sciences certainly appeared to protect 'all and only' humans as a distinct domain of inquiry and concern. Yet, on the other, as we saw in the case of Marx, social science explanations tended towards the naturalistic, whereby humans appeared more as objects than subjects of history. Indeed, this dose of naturalism – what social research textbooks nowadays operationalise as the 'analyst's perspective' – was probably necessary for social scientists to make a clean break from legal theorists and purveyors of commonsense epistemology. Thus, by the end of the nineteenth century, the combined efforts of historical linguistics and empirical anthropology had consigned the idea of an original social contract to a category of myth already inhabited by biblical covenants. Nietzsche's 'genealogy of morals', which treats ancient texts as symptomatic of thoughts and feelings continuous with our animal nature, merely vulgarised what was quickly becoming the dominant trend in social sciences, especially once Darwin's *Origin of Species* and *Descent of Man* entered the scene. The fact that such founders of academic sociology in Germany – and younger contemporaries of Nietzsche – as Ferdinand Toennies and Max Weber continued to stress the contractarian nature of modern society should be understood as an attempt to swim against this current that was driving towards a renaturalisation of the human (Proctor 1991: chap 8). The sociologists proved more-or-less successful for most of the twentieth century, as the ultimate *universitas*, the nation-state, remained the standard-bearer for 'society'. But there is little doubt that their force is dissipating in the face of 'the postmodern condition' (e.g., Urry 2000).

Twentieth century philosophy charted the path to renaturalisation somewhat differently but largely to the same effect. Some combination of Kant and Bentham, especially John Stuart Mill's rule utilitarianism, became quite influential in disciplines with a strong normative top-down approach to society such as constitutional law, public administration and welfare economics. Rawls (1971) was arguably the last great philosophical moment in that tradition. However, in the aftermath of the First World War, theories of 'mass society' and 'complex democracy' cast increasing doubt on the general philosophical ability to harmonize clashing value-orientations within a single ethical system. Logical positivism's response to this problem set the framework for what followed – namely, to treat ethics as a purely formal discipline concerned with moral language (i.e., 'metaethics'), the content of which is determined by the ideologies of the day. Alasdair MacIntyre (1981) dealt a lethal blow to this approach by arguing that ethics becomes irrational if it is not rooted in the life circumstances of those who are called to live by its principles. He embedded this point in a familiar historical account of modernity as the disenchantment of the world but then gave it a distinctly negative spin. Whereas modernists invoked Kant and Bentham as preserving the best of

Christianity without the dubious theological scaffolding, MacIntyre argued that the modernists only succeeded in mouthing rules and principles that in practice did little to constrain people's actual behaviour; hence, the ease with which modernity had slipped into relativism and nihilism.

However, MacIntyre failed to see that the formal character of modernist ethical horizons might reflect the projected experience of creatures who aspire to divinity, and hence to feel *fully* responsible for their actions. Indeed, Kant's and Bentham's universalism bears the clear marks of Christian origins: the decision-maker's detached divine standpoint combined with an *a priori* egalitarian attitude towards all humans, given the shared divine origin of their souls. Thus, Biblical parables that would have us seek God in the most vile and wretched of our fellows are devices to instil a moral sensibility that countermands our 'natural' response to encounters with particular people, which might be revulsion or sheer avoidance. Legal doctrines of 'procedural justice' that instruct officers of the court to undergo heroic abstraction from the specifics of individual lives function in much the same way, a point that Rawls (1971) especially exploited to great effect.

Thus, the fact that Kant and Bentham appeared to presume that people have much greater knowledge *ex ante* and much greater tolerance *ex post* of the consequences of their decisions than is normally expected in everyday life simply shows that they were trying to raise our normative standards, so that humans cannot be confused with mere animals. Unlike MacIntyre's followers in 'virtue ethics', Kant and Bentham regarded our spontaneous moral sensibilities as more raw material than natural norm. Each in his rather different way would have us develop into creatures who can view, say, the suffering of others as symptomatic of a global disorder that needs to be addressed at the level of principle, which is to say, *not* in terms of our spontaneous response to suffering. Thus, we would want to learn as much as possible from the suffering of others to prevent its future occurrence but without necessarily committing ourselves to relieving the suffering of those who have stimulated this change in course of action.

In contrast to this entire line of reasoning, virtue theorists such as Martha Nussbaum trust our spontaneous moral sensibilities, which they then try to ground in 'human nature'. Unfortunately, they follow MacIntyre's lead in relying on Aristotle, sometimes as filtered through Aquinas, when in fact the biological basis for human nature has shifted to Darwin, with the emphasis now on the 'nature' rather than the 'human' part of human nature. As a result, virtue ethics, perhaps unwittingly, has contributed to the renaturalisation of the human, with 'virtue' often nowadays functioning as a quaint word for gene expression.

A good way to see this point is by considering 'reciprocal altruism', the principle that evolutionary psychology nowadays postulates as the foundation of ethics, both human and animal (Wilson 2004). This principle supposes that organisms will do what they can to maximize the survival of their common genetic material. It may effectively mean that some organisms sacrifice themselves for others that are better positioned to reproduce their shared genes. Reciprocity may cross species boundaries, given the vast overlap in genetic makeup amongst life-forms. And while the logic behind reciprocal altruism is sufficiently abstract to be easily treated as a version of game theory, it is meant to apply to specific ecologies consisting of well-defined sets of interacting organisms of various species.

This very embodied and embedded sense of mutual concern recalls the spontaneous displays of 'benevolence' that Scottish Enlightenment figures like Adam Smith and David Hume took to be the natural ground of moral judgement. But it stops well short of the universalistic aspirations of a Kant or a Bentham, who would have us act in a way that treats everyone equally, regardless of the personal impact that someone's existence might have on us. Here it is worth recalling, as Ernst Mayr (1982) famously put it, the ontology of species underlying evolutionary thought is one of sustainable populations rather than categorical types. Although Mayr's point is normally treated as pertaining to the nature of biological science, it equally applies to the moral horizons informed by this science. Perhaps the only significant difference between today's evolutionary psychology and the Scottish Enlightenment is that the former tends to stress the *limits* of our benevolence – namely, it fails to encompass those members of our own species with whom we are unlikely to have any evolutionarily relevant contact, notwithstanding our species-typological likeness to them; hence, our persistent deafness to calls to end poverty in areas far from our homes (Slovic 2007).

In contrast, if we spontaneously thought of humans in more species-typological terms, such that we regarded each person abstractly as an instance of the same type to which we ourselves belong, then our practice would conform to the universalist ideals of Kant or Bentham. Rather than thinking of ourselves as we would most naturally think of others, namely, victims of their fate, we would see them as we naturally see ourselves, namely, capable of taking full responsibility for their actions. This difference between appreciating someone else's plight at a distance and imagining them as virtual versions of ourselves marks a still underestimated distinction between, respectively, *sympathy* and *empathy* (Stueber 2006). One of Christianity's great gifts to epistemology has been its insistence that humans are not simply passive repositories of experience but agents empowered to organize their experience according their own designs. It is not enough to sympathise with where our fellows are coming from. We must also learn to empathise by imagining that, under other circumstances, they could be the ones judging us.

The fugitive essence of 'the human': Towards humanity 2.0?

In this chapter, I have traced the very idea of a domain concerned with 'all and only' humans – the twin root of sociology and socialism in Comte

and Marx and, more generally, social science and democracy – to a specific dynamic conception of *Homo sapiens*. It extends from theological
innovations introduced by John Duns Scotus that collectively licensed
direct comparisons between human and divine being. Although generally
unappreciated by secular philosophers, this genealogy has not escaped the
notice of the recent 'radical orthodoxy' within Anglican Christian theology, which treats the Scotist critique of Aquinas as the founding moment
of modernity, a version of Original Sin in secular time (Milbank 1990). I
largely agree – but minus the negative spin! Scotus diagnosed our difficulty
in comprehending God's infinitude in terms of our failure to see how all
the virtues of mind and body – which in everyday life are invested in quite
different acts, people, and so on – could be equally and ultimately instan-
tiated in God. This idea became the source of the *asymptotic imagination*,
which projects that we get closer to, if not outright merge, with the deity by
increasing and concentrating the virtues, even at the expense of radically
reconfiguring not only ourselves and our social milieu but also our very
material nature.

In this context, Jesus has exemplified how all the virtues might be temporarily consolidated in a single member of *Homo sapiens*. This was certainly
how Joseph Priestley interpreted the life of Jesus, and in the eighteenth and
nineteenth centuries Newton was sometimes portrayed as the 'second coming' of Jesus, given his own remarkable powers of synthesis (Fara 2002).
More mainstream Christians have tended to dismiss such 'Unitarian' interpretations for their apparent denial of the divinity of Jesus. But it would
be more correct to say that Unitarians merely deny the *uniqueness* of his
divinity. In other words, Jesus enabled others to lead similar lives in their
own times, which is something quite different from treating his words as
dictations to subordinates. Of course, that still leaves open the equally – if
not more – heretical suggestion that Jesus did not quite manage to embody
all the divine virtues in his own being, thereby motivating the persistent
drive to self-improvement in secular time.

However, generally speaking, the nineteenth and twentieth centuries were
driven by a more collective if not outright corporeal sense of asymptotically
convergent virtues. The theological roots reach back to John Calvin's interpretation of the Christian sacrament of the Eucharist, whereby Christian com-
municants are said to partake of Christ by ingesting bread and wine. Calvin
took literally the idea that Christ is embodied in any community assembled
in His name for such a purpose. Two centuries later, that heretical native
of Calvin's Geneva, Jean-Jacques Rousseau, secularised Calvin's reading of
the Eucharist as the 'general will', thereby blurring the individual-collective
distinction altogether by defining a normatively acceptable person – what
after the French Revolution was called a 'citizen' – in terms of one's willingness to represent the social whole. This move launched an era of 'ideological' politics, whereby people were routinely expected to stand for ideas,
even if it meant dying for them in the process. We call the extreme end of

this development 'totalitarianism'. Nevertheless, the search for appropriate vehicles of meaning remains.

Nowadays, another two centuries after Rousseau, we have begun to take the asymptotic imagination one step further by seeking a divine convergence of virtues that challenges humanity's biological integrity through the removal of the organism-environment distinction. This move is epitomised by concepts ranging from 'extended phenotype' (Dawkins 1983) to 'cyborganization' (Mazlish 1993), depending on whether one stresses the general evolutionary or more specifically human sides of this process. In either case, to recall Marshall McLuhan's (1964) classic definition of 'media', this project is dedicated to 'the extensions of man [*sic*]' (cf. Fuller 2007a, chap.6), or what, in more religious terms, would be called 'self-transcendence'. As these successive reconfigurations of mind and matter have made clear, the Scotist doctrine left open a profound question: *Does our capacity to come closer to God amount to second-guessing him or, to coin a phrase, 'second-powering' him?*

The two possibilities can be distinguished in terms of what we as creatures *in imago dei* are meant to reproduce of the deity. Here it is worth noting that the classical logical expression of natural laws as universalised conditional statements (i.e., 'For all x, If x is A, then x is B') is ambiguous between the 'second-guessing' and 'second-powering' senses of 'coming closer to God'. An epistemic reading would construe natural laws in 'second guessing' terms as efficient procedures for calculating all the states of the physical universe, given initial conditions. An ontological reading would regard natural laws in 'second powering' terms as the set of parameters within which God brought about those states but could have equally brought about other states (i.e., had 'x' *not* been 'A') – and might do so in the future, perhaps with our help or by what are commonly called 'miracles'. But to appreciate the stakes, the matter is best considered in metaphysical terms.

On the one hand, to second-guess God would be to figure out his plan and manifest it as best we can. A secular residue of this view is the idea that the ultimate scientific theory should provide a true and complete representation of all reality. This is what philosophers call 'convergent scientific realism', a position that perhaps originated with Priestley but is most closely associated with the pragmatist Charles Sanders Peirce, an avid fan of Duns Scotus (Laudan 1981, chap. 14). It seems to imply that a point will come when the work of God (and humanity) is done – in other words, that there is an 'end of science'. On the other hand, to second-power God would be to reproduce the sphere of possibilities from which the deity selected a particular plan to realize, so as to maintain the deity's creative freedom. Among the secular residues of this view is the frequent association of technological progress with an increasing ability to keep options open, often by reversing or mitigating the bad effects of earlier choices. In this context, Popper turns out to be God's best friend.

However, as social science has drifted from its theological moorings, the terms of democracy are increasingly up for renegotiation. On the one hand, normative categories traditionally confined to humans, especially 'rights', are being extended to animals and even machines. On the other hand, there are increasing attempts to withhold or attenuate the application of these categories to the disabled (including the virtually disabled, via antenatal screening), the elderly, the chronically ill or simply unwanted or unproductive humans. This characteristically postmodern 'open borders' approach to life is often presented in a positive light as a triumph for anti-essentialism. However, essentialism about 'the human' has arguably always been about using ontology to stiffen moral and political resolve. It is difficult, if not impossible, to define a system of complementary rights and duties that might constitute 'human dignity' without a clear sense of the range of beings eligible for the title 'human'.

From this standpoint, the confidence with which self-styled 'critical' theorists continue to claim to be able to demarcate the human from the nonhuman rings hollow. For example, both Jürgen Habermas and Noam Chomsky fall back on language as the defining human trait. But Habermas (1981) does little more than democratize Aristotle's account of the human as *zoon politikon* by extending the intersubjectivity displayed in the public sphere to cover all of social life. In effect, and perhaps in keeping with modern sensibilities, he redefines much of the private as public, which then enables him to include all of *homo sapiens* as properly 'human'. As for Chomsky's appeal to language, it is more sophisticated, based on its endlessly creative character – at least when understood at the syntactic level. This last qualification is important because, over the years, Chomsky has increasingly restricted language's creativity to the sheer combinatorial powers of its grammatical units. Conceding recent research in evolutionary psychology, he has even felt the need to acknowledge that many higher apes are capable of complex forms of signification, though *Homo sapiens* remains unique in its ability to combine signs and build upon them, especially in a self-reflective fashion – at least until evolutionary psychology shows otherwise (Hauser, Chomsky, Fitch 2002).

Even if Foucault overstated the case that humanity is receding as a salient object on the metaphysical horizon, it would appear that humanity's nature is forever fugitive. Any proposed 'mark of the human' has been the target of simulation (usually by machines) and extension (usually to animals), the overall effect of which has been to cast doubt on the original intuitive applications of 'the human' to members of *Homo sapiens*. Just as computers potentially outstrip us in the capacity for reasoning and information processing, animals increasingly appear more complex and adaptive to nature. In that case, from both a scientific and a political standpoint, what is worth continuing to defend as distinctly 'human'? Perhaps the most ambitious strategy for projecting the predicate 'human' into the twenty-first century has centred on 'converging technologies' ('CT') – that is, the integration of cutting-edge research in nano-, bio-, info- and cogno-sciences for purposes of extending the power and control of human beings over their own bodies and their environments. CT, presented with interesting variations, is now a part of the long-term science policy agendas of all the major nations but its roots reach back to Francis Bacon's scientific commonwealth and Count Saint Simon's utopian socialism (Fuller 2009). There are at least six variants of CT, each of which may be associated with the sense in which it would have 'the human' projected:

1. *Humanity Transcended:* Julian Huxley's original sense of 'transhumanism', namely, the return of natural selection to its metaphorical roots in artificial selection, such that humans become the engineers of evolution. One might see this as 'reflexive evolution'. Huxley originally conceived it along classically eugenicist lines, whereby the state would be empowered to make strategic interventions to encourage or discourage the frequency of various traits. In cases of doubt as to the exact genetic component of traits, one would remove social barriers so as to allow 'equal opportunity' (a Huxley coinage) for gene expression. Advances in knowledge of genetics allow these interventions to become increasingly upstream (e.g., antenatal screening), which preempt traditional forms of social engineering, such that eventually we acquire Plato-like powers to legislate the desirable distribution of traits in a population, which in principle may be quite different from the traits' normal frequency.

2. *Humanity Enhanced:* Most CT science policy statements focus on the prospect of humans acquiring improved versions of their current powers without the more extreme implications of prolonging life indefinitely or upgrading us to a superior species. For example, the use of nanotechnology to eliminate fatty deposits from arteries or clean polluted water is designed simply to raise people's productivity and quality of life, so that they can contribute more efficiently to the economy and the ecology in their normal life spans.

3. *Humanity Prolonged:* This CT goes beyond enhancing normal life capacities and towards suspending, if not reversing, age-related disintegration and perhaps even death itself. This aspiration brings together theological (both Greek and Christian) preoccupations with immortality and modern medical science's conception of death as the ultimate enemy that needs to be overcome. It is specifically focussed on extending indefinitely human existence in its prime and hence explicitly raises questions about intergenerational fairness, if not the very need for intergenerational replacement or, indeed, sexual reproduction itself.

4. *Humanity Translated*: This CT is arguably a high-tech realization of theological ideas concerning resurrection, whereby an individual's distinctive features are terminated in one physical form and reproduced in another.

It is usually discussed in terms of the uploading of mental life from carbon- to silicon-based vehicles, typically with the implication that the relevant human qualities will be at once prolonged, enhanced and transcended. But it may also include avatars ('second lives') in virtual reality domains, whereby the individual's human existence becomes coextensive with participation in this posthuman translation process.

5. *Humanity Incorporated*: The take-off point for this CT is the legal category of artificial persons and corporate personalities, the medieval artifice of the *universitas*, and includes all of its premodern, modern and postmodern affiliates: the extended phenotype, the 'supersized mind' of 'smart environments' (Clark 2008) and, most radically, 'systems architecture' (Armstrong 2009), whereby human and nonhuman elements are not only combined but allowed to co-develop into novel unities. All of these proposals share the idea that humanity's distinctiveness comes from our superior organic capacity to make the environment part of ourselves. As Hobbes had already suggested in *Leviathan*, it is an especially materialist take on the *imago dei* doctrine.

6. *Humanity Tested:* In light of CT's quite speculative character, the enthusiasm with which large sectors of the scientific and political communities would embrace such an agenda – and not for the first time – speaks to humanity's arguably superior capacity for experimenting with one's own life and the lives of others. Reflecting the likely, perhaps even disastrous, failure of many of the CT experiments involved in realizing any of the previous five projections of humanity, the focus here is on promoting a culture tolerant of risk-taking, say, by a generous social insurance scheme, a supportive environment for coping with unanticipated outcomes and a strong sense of an overarching long-term collective project.

Finally, in light of these multiple projections of the human future, how should the history of the concept of humanity inform the future of social science? The guiding insight that connects the theological backdrop of the Scientific Revolution and today's convergent technologies agenda is the idea that biology is a branch of technology, that is, the application (by whom?) of physical principles for specific ends. The theological ground for this interpretation is a rather literal understanding that humans have been uniquely created in the image and likeness of the ultimate artificer of nature. This provides *prima facie* license for the reverse engineering and enhancement of our animal natures (Harris 2007). Ontologically speaking, we can be 'creative' just as God can, though perhaps not to such great effect. However, I say '*prima facie* license' because there may also be epistemic, ethical and political grounds on which to regulate, if not curtail, our godlike powers. For example, we might query the degree of risk at which particular individuals or entire populations are placed by biotechnological interventions or, as Habermas (2002) has emphasized, the degree of 'dehumanisation' as conventionally understood that would result, were such interventions to issue in desirable outcomes on a reliable basis.

But even as these normative matters remain unresolved, we need to be clear which objections to the idea of 'Humanity 2.0', so to speak, are merely epistemic and which truly ethical or political. This distinction is often obscured by preemptive ontological arguments about the 'limits' of human nature. There is no doubt that humans are improving their capacity to manipulate and transform the material character of their being. Of course, uncertainties and risks remain, but in the first instance, they are about how costs and benefits are distributed in this imperfect process – that is, unless one has a principled objection to humans being changed in certain ways. But principled objections do not require an ontological basis. Here critics need to catch up with the times. Traditional natural law appeals to 'violations of human nature' of the sort frequently invoked by George W. Bush's bio-ethics panel increasingly lack intellectual currency, given the socially constructed character of humanity and the anti-essentialism of modern biology (cf. Baillie and Casey 2004). Instead, 'principled objections' to Humanity 2.0 should follow the example of resisters of technological innovations, namely, to argue that unregulated innovation is likely to increase already existing inequalities in society. In other words, an explicit policy of redistribution, which in turn addresses both the external and internal boundaries of 'our' society – that is, who *prima facie* counts as human and to which category of human they most relevantly belong – always needs to attend the introduction of any technology capable of radically redefining the human field.

References

Armstrong, R. (2009) 'Protocells and Plectic Systems Architecture.' *Sophia* 3 (June): 17. http://www.ucl.ac.uk/~ucbpeal/sophia/issue3-web.pdf

Baber, Z., ed. (2009) 'Review Symposium on Steve Fuller's *The New Sociological Imagination.' History of the Human Sciences* 22, 110–45.

Baillie, H. and T. Casey, eds (2004) *Is Human Nature Obsolete?: Genetics, Bioengineering, and the Future of the Human Condition.* Cambridge, MA: MIT Press.

Brague, R. (2007) *The Law of God: The Philosophical History of an Idea.* Chicago: University of Chicago Press.

Brams, S. (2002) *Biblical Games: Game Theory and the Hebrew Bible.* Cambridge, MA: MIT Press.

Clark, A. (2008) *Supersizing the Mind.* Cambridge, MA: MIT Press.

Corbey, R. (2005) *The Metaphysics of Apes.* Cambridge: Cambridge University Press.

Dawkins, R. (1983) *The Extended Phenotype.* Oxford: Oxford University Press.

Dawkins, R. (2006) *The God Delusion.* New York: Houghton Mifflin.

Deleuze, G. and F. Guattari (1987) *A Thousand Plateaus.* (Orig. 1980) Minneapolis: University of Minnesota Press.

Fara, P. (2002) *Newton: The Making of Genius.* New York: Columbia University Press.

Fleischacker, S. (2004) *A Short History of Distributive Justice.* Cambridge, MA: Harvard University Press.

Foucault, M. (1970) *The Order of Things: An Archaeology of the Human Sciences.* New York: Pantheon Books.

Fuller, S. (1997) *Science.* Milton Keynes, UK: Open University Press.

Fuller, S. (2002) 'Making Up the Past: A Response to Sharrock and Leudar.' *History of the Human Sciences* 15(4), 115–23.

Fuller, S. (2003) *Kuhn vs. Popper: The Struggle for the Soul of Science.* Cambridge: Icon.

Fuller, S. (2006) *The New Sociological Imagination.* London: Sage.

Fuller, S. (2007a) *New Frontiers in Science and Technology Studies.* Cambridge: Polity.

Fuller, S. (2007b) *Science vs. Religion? Intelligent Design and the Problem of Evolution.* Cambridge: Polity.

Fuller, S. (2008) *Dissent over Descent: Intelligent Design's Challenge to Darwinism.* Cambridge: Icon.

Fuller, S. (2009) 'Knowledge Politics and New Converging Technologies.' *Innovation* 22(1), 7–34.

Funkenstein, A. (1986) *Theology and the Scientific Imagination.* Cambridge: Cambridge University Press.

Goodman, N. (1954) *Fact, Fiction, and Forecast.* Cambridge, MA: Harvard University Press.

Gray, J. (2002) *Straw Dogs: Thoughts on Humans and Other Animals.* London: Granta Books.

Habermas, J. (1981) *A Theory of Communicative Action.* Boston: Beacon Press.

Habermas, J. (2002) *The Future of Human Nature.* Cambridge: Polity.

Harris, J. (2007) *Enhancing Evolution.* Princeton, NJ: Princeton University Press.

Harrison, P. (1998) *The Bible, Protestantism and the Rise of Natural Science.* Cambridge: Cambridge University Press.

Hauser, M., N. Chomsky, and W. Fitch (2002) 'The Faculty of Language: What Is It, Who Has It, and How Did It Evolve?' *Science* 298, 1569–79.

Heims, S. (1991) *Constructing a Social Science for Postwar America: The Cybernetics Group, 1946–1953.* Cambridge, MA: MIT Press.

Ingold, T. (1994) 'Humanity and Animality'. In: *Companion Encyclopedia of Anthropology*, ed. T. Ingold. London: Routledge, chap. 2.

Israel, J. (2001) *Radical Enlightenment.* Oxford: Oxford University Press.

Johnson, S. (2008) *The Invention of Air: A Story of Science, Faith, Revolution, and the Birth of America.* New York: Penguin.

Koerner, L. (1999) *Linnaeus: Nature and Nation.* Cambridge, MA: Harvard University Press.

Kurzweil, R. (1999) *The Age of Spiritual Machines.* New York: Random House.

Latour, B. (1993) *We Have Never Been Modern.* Cambridge, MA: Harvard University Press.

Laudan, L. (1981) *Science and Hypothesis*: Dordrecht, Netherlands: Kluwer.

MacIntyre, A. (1981) *After Virtue.* South Bend, IN: Notre Dame Press.

Margalit, A. (2004) *The Ethics of Memory.* Cambridge, MA: Harvard University Press.

Mayr, E. (1982) *The Growth of Biological Thought.* Cambridge, MA: Harvard University Press.

Mazlish, B. (1993) *The Fourth Discontinuity: The Co-Evolution of Humans and Machines.* New Haven, CT: Yale University Press.

McLuhan, M. (1964) *Understanding Media.* New York: McGraw-Hill.

Murray, C. (2003) *Human Accomplishment.* New York: HarperCollins.

Nadler, S. (2008) *The Best of All Possible Worlds.* New York: Farrar Straus Giroux.

Nagel, T. (1986) *The View from Nowhere.* Oxford: Oxford University Press.

Noble, D. (1997) *The Religion of Technology*. New York: Alfred Knopf.

Passmore, J. (1970) *The Perfectibility of Man*. London: Duckworth.

Popper, K. (1957) *The Poverty of Historicism*. New York: Harper and Row.

Proctor, R. (1991). *Value-Free Science?* Cambridge, MA: Harvard University Press.

Rawls, J. (1971) *A Theory of Justice*. Cambridge, MA: Harvard University Press.

Schneewind, J. (1984) 'The Divine Corporation and the History of Ethics.' In: *Philosophy in History*, ed. R. Rorty, J. Schneewind and Q. Skinner. Cambridge: Cambridge University Press, pp. 173–92.

Schneewind, J. (1997). *The Idea of Autonomy*. Cambridge: Cambridge University Press.

Singer, P. (1999). *A Darwinian Left*. London: Weidenfeld & Nicolson.

Slovic, P. (2007). 'Genocide: When Compassion Fails.' *New Scientist* 2598 (07 April): 18.

Stueber, K. (2006). *Rediscovering Empathy: Agency, Folk Psychology, and the Human Sciences*. Cambridge, MA: MIT Press.

Toulmin, S. (1990) *Cosmopolis: The Hidden Agenda of Modernity*. New York: Free Press.

Urry, J. (2000) *Sociology beyond Societies*. London: Routledge.

Wallerstein, I., et al. (1996) *Open the Social Sciences*. Palo Alto, CA: Stanford University Press.

Williams, T. (2007) 'John Duns Scotus.' In: *Stanford Encyclopedia of Philosophy*, ed. E. Zalta, http://plato.stanford.edu/entries/duns-scotus/. Accessed 24 March 2009.

Wilson, C. (2004) *Moral Animals: Ideals and Constraints*. Oxford: Oxford University Press.

Wilson, E. O. (1998) *Consilience: The Unity of Knowledge*. New York: Alfred Knopf.

Index